To Diana

from.

Teddy and Berney

July 1974

Diana de C-Wheeler.

Drummin.

Carbury.

Co. Kildare

Eire.

WILLINGLY TO SCHOOL
How Animals are Taught

WILLINGLY TO SCHOOL
How Animals are Taught

BY

Hermann Dembeck

TRANSLATED BY

CHARLES JOHNSON

George G. Harrap & Co. Ltd
London · Toronto · Wellington · Sydney

First published in Great Britain 1970
by GEORGE G. HARRAP *&* CO. LTD
182 High Holborn, London, W.C.1

SBN 245 59982 7

Printed in Great Britain by
Western Printing Services Ltd, Bristol
Made in Great Britain

PREFACE

BECAUSE of some special faculty or skill or talent, certain species of animals seem disposed to learn a variety of tricks which are often contrary to their way of life in the wild. Take for instance the spirit of friendliness, even devotion, which a competent animal trainer can instil into even the largest members of the cat family. Whether it is man or the 'wild' animal that contributes most is a moot point; basically, it is a matter of team-work. The results can, of course, sometimes be negative, though generally it will be found that even lions are receptive to the art of gentle persuasion. Time and again, during my thirty years' close and on the whole friendly association with animals, I have confirmed the truth of this.

To test a pet theory of mine, I went for a stroll one day with a fully grown yet completely tame tiger. So long as either the trainer or I was near, all was well. But when our backs were turned the tiger clawed a keeper through the bars of its cage. The man had evidently failed to gain the animal's friendship. Perhaps he had set about it in the wrong way; perhaps, in spite of his considerable experience with 'big cats', he had incurred an intense dislike on the part of the tiger—a common enough occurrence among human beings.

Some time later a fully grown lion accompanied me without incident for a whole summer, until it was maltreated by someone, whereupon it changed its whole attitude towards people. It ceased to regard them as 'reliable friends' and, though it still tolerated them, it no longer gave them its complete confidence.

The tame elephants I have known have included some willing to work and just as willing to behave mischievously. Anthropoid apes revealed themselves to be clever, skilful artists and useful helpers on the one hand, but unpredictable creatures on the other, having a good deal more in common with other animals than with man; in them one sees most clearly the limitations of aptitude and the restricted extent to which animals can be trained. The teachability—or perhaps it is merely the imitative instinct or obedience—of many animals is known to all of us, either through our own dealings with them or through visits to the circus. However, it is only within the last few years that the forgotten talents of dolphins have been quite unexpectedly rediscovered, though in fact they were noted in ancient times. Modern zoologists and animal trainers have repeated experiments relating to the teachability of animals described in old writings and long regarded as dubious legends, only to find that these ancient chroniclers were substantially accurate.

There are frequent references in the Bible to natural—if unusual—talents possessed by various creatures: a dove released by Noah returns to the Ark with an olive-leaf in its beak; Balaam's ass speaks. Neither of these was in any way trained; what they performed was symbolic, miraculous. But from these stories it becomes clear that men had already invested animals with the ability to be obedient, even rational, to an unusual degree. Other records—ancient Egyptian fragments of papyri, pictures, cuneiform inscriptions and colour reliefs discovered in the Mesopotamian region—reveal that ancient man was aware of the diverse aptitudes not only of different species of animals but also of individual creatures endowed with special intelligence, and that he made efforts to train them.

Although the training methods practised in the past by travelling entertainers were torment for the animals, a crime against the basic rights of all living creatures, and swindling of the public, 'training by intimidation' has not been in common practice now for ages. The basis of a successful training method involves, rather, complete trust and understanding between man and beast, aided by the animal's play instinct and man's ability to channel that instinct by empathy and scientific knowledge. It is with this kind of training that this book is concerned.

HERMANN DEMBECK

CONTENTS

PART V: THE FRIENDLY DOLPHINS

ILLUSTRATIONS

Illustrations

TRANSLATOR'S NOTE

Mr and Mrs John Elphick read my translation and made many helpful suggestions, nearly all of which I have accepted with gratitude.

PART ONE

THE ARTFUL MONKEYS

CHAPTER ONE

GUARDIANS AND PETS OF ANCIENT PEOPLES

It will surprise no-one to know that apes and monkeys occupy first place in the intelligence rating of the animal world and are generally agreed to be the cleverest of those animals which can be trained. In spite of this there is not such a close bond now between them and human beings as there used to be. People love to see them at the zoo, where their antics and pranks are an endless source of amusement, or in the circus ring, where their obedience and intelligence are admired. But hardly anyone nowadays thinks of keeping a monkey as a house pet. This was not always so; thousands of years ago monkeys were regarded as sacred, and there are many accounts of their being spoilt, pampered, dressed up, and treated almost like children.

There is evidence that four thousand years ago the Egyptians kept sacred baboons in their homes and temples. These hamadryas baboons were sacred to Thoth, the ibis-headed god of the moon and patron of scholarship. The numerous civil servants employed by the Pharaohs and regional governors considered themselves equal in status to the scholars, and so adopted the baboon as their own patron. Thus the baboon came to be worshipped as the tutelary god of civil servants and judges.

Even today carved monkeys are commonly found on the desks of senior civil servants. These represent either sacred baboons in their role as tutelary gods or three rhesus monkeys sitting in a row, the first holding his hands over his eyes, the second over his mouth, and the third over his ears—an Indian symbol of wisdom: see no evil, speak no evil, hear no evil. Perhaps for people in high office this trio of monkeys serves as a reminder that however hectic their jobs may be they should keep a sense of balance and detachment. The reason why baboons in particular became the patrons of civil servants was chiefly that they are so adaptable.

In ancient Egypt 'sacred animals' were kept in the temple grounds. Baboons lived in cages, the Apis bull in a stall, crocodiles in specially constructed ponds. The temple cats were allowed to roam freely in the priests' quarters. The Pharaohs were allowed to keep baboons in their palaces because, as well as holding supreme temporal power, they were initiated in the most secret sacerdotal rites. They were priest-kings, as were Montezuma among the Aztecs and Atahualpa among the Incas, thousands of years later.

These highly prized monkeys originated from the land of Punt, also known as Troglodyte Land. Punt is identical with modern Somalia. In those days it produced ivory, valuable timbers, myrrh and incense, gold, and precious stones. There is an account of a particularly successful expedition, undertaken about 2400 B.C. on behalf of the second Pharaoh of the Fifth Dynasty, in J. H. Breasted's *Ancient Records of Egypt*. How active Punt trading was in those days is shown by an epitaph dating from the Sixth Dynasty, inscribed for a helmsman named Khnemhotep who lived in Elephantine, declaring that he had sailed to Punt with his captain no fewer than eleven times.

The monkeys brought back from Punt, after voyages lasting several months, were so expensive that only wealthy dignitaries could afford them. The Pharaohs possessed the sole right of disposal of any animals brought back on vessels commissioned by them. They generally presented the monkeys to their most senior state officials—who were obliged to show their gratitude by giving valuable gifts in return! The upkeep of these monkeys must also have been considerable, as suitable food could not have been easy to obtain.

One of these well-to-do animal lovers was Nebewechmet, a

court steward to the Pharaoh Chephren, who lived about 2700 B.C. He owned two large, tame, long-haired sacred baboons. They were males, but by all accounts very sociable. Nebewechmet and his wife would take both of them along when they went on a tour of inspection of their employees. "To be sure, the great lord took extreme delight in the inspection which, on such occasions, the monkeys made of his workmen" (Adolf Ermann).

Pictures from Egypt of various long-tailed monkeys show these too only in 'upper-class' surroundings. The monkeys are either depicted squatting under easy-chairs of such superior quality that they could only have belonged to the wealthier members of society, or painted on costly tombs. Like the baboons, these monkeys had a much better life than many of the peasants. Noble ladies would painstakingly dress the monkeys' hair and adorn them with arm and foot bangles, presumably of precious metal; and it was not unusual to put neck ornaments on their pets as well. Egyptologists have proved that long-tailed monkeys and baboons even shared their mistresses' apartments. There they could watch them making up their faces and eyebrows and admiring themselves in the polished metal hand-mirror. Seeing their own reflection there, the monkeys would look behind the mirror in search of the 'other' monkey—as monkeys still do. Surviving pictures give an indication of the wealth of such women as could afford to keep a monkey. A mirror was regarded as an essential item for people of high social standing; the poor maidservant could certainly never hope to own one, and had to look in a pond to see her reflection. Nor could she afford those exquisite rouge-pots and jewel-boxes on which are found monkeys painted by a master hand.

A great deal of information is available concerning an Egyptian expedition to the land of Punt about 3500 years ago. Several hundred men, in five large vessels commissioned by Queen Hatshep-sut, took part in it; a hundred and fifty oarsmen were needed to propel the massive galleys forward when they were becalmed.

This trading expedition, in 1493 B.C., was the first to be undertaken following Egypt's liberation from the yoke of the Hyksos and, though there had been no contact for centuries between

Egypt and Punt, the Egyptian Queen's envoy and his companions were hospitably received by the ruler of Punt. Inscriptions in the temple of Dair al-bahri, the mausoleum of Queen Hat-shep-sut, which may still be read, bear witness to the success of the expedition:

> The vessels were fully laden with the marvellous produce of the land of Punt—many kinds of costly timbers, great store of fragrant resin and fresh incense, ebony, ivory set in pure gold from the land of Aamu, sweet-smelling kesit-wood, ahem-incense, sacred resin painted to delight the eye, dog-headed and long-tailed monkeys, greyhounds, leopard pelts, and native inhabitants together with their children.

This temple also has inscriptions and drawings which explain how monkeys were transported to Egypt. Even before the ships set sail from the Troglodyte coast the monkeys were allowed complete freedom of movement. They clambered up the rigging onto masts and yards, where they could frolic as they pleased. When they got hungry or tired they climbed down to their crates, in which food had been placed ready for them.

In the illustrations of the vessels of Queen Hat-shep-sut several baboons are shown on the upper deck next to bales and sacks, some sitting, others running about, and all so clearly recognizable that they could well have been sketched by a modern artist with an eye for anatomical accuracy. One of them is seen being enticed with a piece of fruit by a scribe perched on top of the after-cabin.

The drawings confirm that the inhabitants of the Somali coast had themselves tamed baboons. The ship is still moored to the shore. Unless they had been completely tame and harmless, the Egyptians would hardly have allowed the baboons to run about on board with such freedom. The hieroglyphic text states that the illustration shows the loading of ships on the coast of Punt.

Although there is no historical evidence to support the theory, it is generally assumed that on arrival in Egypt the monkeys and baboons were transported overland in cages on a five-day trek from the port of Leukos Limen, now called Kosseir, along the old caravan route to Coptos on the Nile, near Thebes, where they were handed over to trainers to be looked after and trained for Ammon's temple and the Queen's palace.

It seems likely, from pictures and inscriptions in the temple of Dair al-bahri, that the ruler of Punt and his plump wife accompanied the party to Thebes, and it can just as safely be assumed that the dark-skinned royal couple made the 1250 miles' journey home with the same Egyptian fleet, when next it sailed, laden with gifts presented in return for theirs. From then on, regular sea trading must have been resumed. The numerous sculptures of baboons made in the succeeding centuries prove that many apes and monkeys were brought from Punt to Egypt.

We know from the Bible that King Solomon's merchant fleet also imported apes. The Phoenician commander-in-chief of the navy, Hiram, King of Tyre, sailed by arrangement with Israel to the land of Ophir to barter goods. On these trips he brought several species of monkeys and apes back to Ezion-geber, an Israelite port on the Red Sea.

It is still impossible to state with certainty where Hiram's destination, Ophir, is located—whether it is on the east coast of Africa, or whether Hiram and his mariners knew a sea route to India. The reference in the Bible is imprecise about this:

> And king Solomon made a navy of ships in Ezion-geber, which is beside Eloth, on the shore of the Red sea, in the land of Edom.
>
> And Hiram sent in the navy his servants, shipmen that had knowledge of the sea, with the servants of Solomon.
>
> And they came to Ophir, and fetched from thence gold, four hundred and twenty talents, and brought it to king Solomon. (1 Kings ix, 26–28.)
>
> For the king had at sea a navy of Tharshish with the navy of Hiram: once in three years came the navy of Tharshish, bringing gold, and silver, ivory, and apes, and peacocks. (1 Kings x, 22.)

Solomon's old shipyard, the biblical Ezion-geber, was discovered by American archaeologists in 1940 beneath the rubble of a hill called Tell-el-Keleifeh. And from the port of Aqaba, not far from the old Ezion-geber, Israeli ships have sailed in modern times through the Red Sea. Werner Keller quotes, in his book *The Bible is right*, a Phoenician priest named Sanchuniathon who says of King Solomon's fleet: "... and Joram [*i.e.*, Hiram] found himself obliged to have the wood taken there on 8000 camels. From this a fleet of ten ships was built." Keller continues: "Even the names of the Phoenician captains who had command of the fleet were known to Sanchuniathon. The 'good shipmen' were named Kedarus, Jaminus, and Kotilus." After discussing the theory that Ophir might have been situated on the southern Rhodesian border, where in ancient times gold-mines were worked, Keller comes to the same conclusion as the explorer Albright: "Various clues point, however, to East Africa. This would also be entirely consistent with the time taken for the voyage as specified in the Bible. Furthermore, the bartered goods, such as gold, silver, ivory, and apes, clearly point to Africa as the country of origin." Keller finally identifies the land of Ophir with the ancient Egyptian trading centre Punt, which, as previously mentioned, had been exporting apes and monkeys to Egypt thousands of years before Solomon's time.

Between 530 and 480 B.C. Carthage sought to expand its trade as far as the west coast of Africa, with the intention of setting up new trading stations abroad to obtain local products for sale. With this in view, a fleet of sixty ships was fitted out, about the year 525, to take some 30,000 settlers, craftsmen, and merchants to various points along the coast where harbours could be constructed. Hanno, one of the two highest officials of the city-state of Carthage, was appointed commander-in-chief of the fleet. He wrote an account of this reconnaissance voyage which was later translated into Greek. The original was either taken or destroyed by the Romans when they conquered Carthage. Parts of the Greek translation ultimately reached Heidelberg, and from this *Codex Heidelbergensis 398* we know about the Carthaginians' encounter with anthropoid apes.

At the southernmost point of their voyage, in what is now the Cameroons Bay, the vessels which had not already put in at various other places along the west coast to set up trading posts arrived at a well-wooded peninsula, about which Hanno writes:

> On the far side of the bay there was an island in a lake inhabited by 'men of the woods'. There were, however, a still greater number of women, whose bodies were shaggy, whom our interpreters called Gorillae. We pursued the men, but were unable to catch any, for they all fled, escaping over the

> precipices and defending themselves with stones. We captured three women, but they bit and scratched the leaders and would not follow them. We killed them and flayed them, and brought their skins back to Carthage. We penetrated no further as we were lacking supplies.

The Carthaginians' encounter with the 'men of the woods' has been fixed by Richard Hennig in his *Terrae incognitae* at about 525 B.C.

The Greek historian and statesman Polybius, the close friend and adviser of the Roman conqueror of the Carthaginians, was able to copy Hanno's account prior to the burning of the temple of Baal. Unfortunately of the forty books comprising this history of Rome only five have survived. The loss of the manuscripts has deprived us of further important information about Hanno's 'gorilla adventure'. Pliny and Arrian also wrote briefly about this voyage; it may be that Polybius' *History* was not available in full to them either.

Pliny describes the 'Gorillae' mentioned by Hanno as 'Gorgades', by which term he refers, in the common parlance of those days, to their repulsive appearance and terrifying faces. Other ancient writers have also referred to 'Gorgades' when describing 'men of the woods'—*i.e.*, gorillas. Diodorus used the term 'Gorgons', which he probably took from the monster Gorgon which lived on the sea-coasts. According to the legend, anyone who looked at the hideous face of the Gorgon Medusa was turned to stone with fright. Perseus succeeded in cutting off this monster's head. The gorillas seen by Hanno's sailors must have seemed equally terrifying.

Just what Hanno meant by 'Gorillas' is discussed by Richard Hennig: "That they were not human beings is suggested by the specific emphasis given to the thick covering of hair, for purebred negroes, who alone can be considered in this context, have no hair whatever on any part of their bodies except their heads." Hennig then refers to an expert opinion given by his zoological adviser, Professor Stechow, according to whom "There is no animal which makes use of an implement", in order to show that Hanno's 'Gorillas' could therefore only have been human.

Nowadays we know what in 1935 Professor Stechow could not have known—that South American capuchin monkeys (to cite just one example) use stones to dislodge nuts from trees; we also know that baboons and chimpanzees use sticks and stones, and that in certain situations these implements are also used by gorillas.

The zoologist Dr Knottnerus-Meyer states categorically: "Chimpanzees enjoy beating things with sticks." In his book *Tiere im* Zoo (Animals at the Zoo), dealing with Rome Zoo, of which he was the curator, he writes about a chimpanzee called Max who took great pleasure in beating his companion Virginie with sticks, and another called Sambo, from the South Cameroons, which even attacked human beings with a stick. There is also an account of a female chimpanzee living in the same cage as a pair of young hamadryas baboons, who always beat the male if she could lay hands on a stick.

Dr Knottnerus-Meyer also mentions that a pig-tailed macaque vented its indignation on other monkeys, or on humans, "by throwing everything within reach at them"—unusual with this particular species. He has noted similar behaviour mainly among older baboons, Barbary apes and other macaques. If there are no sticks or stones available the baboons throw sand or even use their feeding dishes as missiles.

When Gargantua, a male gorilla weighing almost forty stone which belonged to the Ringling Circus in the U.S.A. until its death in 1949, got angry with anyone it would hop up and down screaming, pick up its favourite toy, a car tyre, from the floor of the cage and hurl it against the bars, which it would shake so violently that onlookers were terrified the cage would be torn apart. Instead of the American veterinary surgeon Dr J. Y. Henderson and the keeper, both of whom were well known to the gorilla, imagine two Carthaginian warriors on the Cameroons peninsula confronted by the 'wild men of the woods', the gorillas, and substitute a stone for the car tyre! From this observation of Dr Henderson it is clear that gorillas will make use of the nearest available objects as missiles.

The choice of a broom-handle as a weapon has been observed by Dr Henderson even among very much smaller species of monkeys. The male of a pair of Diana monkeys got hold of one and beat his mate with it so viciously that the poor creature sustained serious injuries.

I have myself seen a vervet monkey in Willy Hagenbeck's circus snatch a walking-stick he held out to it and strike a photographer so hard on the arm with it through the bars that he was paralysed for several minutes.

In his book *Beasts and Men* Carl Hagenbeck, the creator of the animal park at Stellingen, near Hamburg, describes the way chimpanzees and orang-utans use stones as missiles. Even less highly developed species of monkeys do the same. Regarding the use of stones as missiles by baboons, Eugene N. Marais, in his book *My Friends the Baboons*, says: "Two males were in the habit of throwing stones at the badgers whenever they had the chance."

Willy Hagenbeck, who trained polar bears, and some colleagues were present when, in 1910, a tame baboon which shared a large training enclosure with seventy polar bears seized a stout stick thrown to it by a spectator and set about the polar bears with it.

I have seen a squirrel monkey, a small, harmless species, use a stone to break open Brazil nuts. On another occasion this same monkey struck out with a small stick at a large dog which was jumping up at its cage.

It could therefore very well have been skins of manlike apes that the Carthaginians took home with them. The 'Holy Mountain' region of the Cameroons is still inhabited by gorillas. There are, it is true, chimpanzees there too. The only question remaining is whether the animals they captured were in fact gorillas or chimpanzees.

The word 'gorilla', as used by Hanno, gives no direct clue as to which of the higher apes he is referring to. But his references to the 'men of the woods' do suggest that he could hardly have been alluding to human beings, since from the very earliest times weapons of various kinds have been in use among negro Africans—bows and arrows, as well as spears. Without doubt, if the attackers had been negroes they would have used such weapons. The word 'gorilla' signified 'scratcher' in neo-Punic, the spoken and written language of Carthage; in the singular the word was written 'gorel', in the feminine plural 'gorillai' and in the feminine singular 'gorilla'.

CHAPTER TWO

ENTERTAINMENTS OF THE GREEKS AND ROMANS

THE demand for small monkeys in ancient times was periodically very great. Carthaginian merchants acquired them in what are now Algeria, Morocco, and Mauretania. They also bought up all the apes they could—and they were numerous—on the island of Pithecusa, known as Monkey Island (the modern Ischia). Presumably there was great interest, even in those days, in the highly gifted Barbary apes, then present in large numbers at the southernmost tip of Spain, but nowadays found only in Gibraltar. The Carthaginians supplied these apes and monkeys to towns in Etruria, Greece, and Rome. In Athens, during the pre-Christian era, monkeys from East Africa were also a common sight, having been brought back by seafarers from ports in the Indian Ocean and transported across Egypt. Among the well-to-do Etruscans and Greeks the smaller long-tailed monkeys were particularly popular as domestic pets. These intelligent monkeys are easily tamed, as every animal lover who has kept them knows.

For a long time it was supposed that various species of long-tailed monkeys occurred in antiquity in North Africa. There is, however, no evidence to support this theory. The import of such monkeys was rather the result of the centuries-long existence of Carthaginian trading posts south of Senegal. It is quite possible that these continued to be operated by the Romans after they had sacked Carthage, and that the Carthaginian sailors worked on in the service of Roman merchants. History tells us nothing definite about this—but then, the activities of merchants seldom figure large in historical records.

As early as the fourth century B.C. monkeys were kept as pets in Greek cities on such a large scale that Eubulos, a comic poet who lived in Athens about 350 B.C., warned women and children not to allow even the tamest monkeys to run about the house unsupervised because they were liable to turn nasty and bite people. Nor was it becoming for ladies, he claimed, to allow tame monkeys to sit next to them at meal-times or to have them on their laps, as though they were small children.

Monkeys were taught to perform various tricks, including riding, somersaults, hand-stands, and apparently even playing the flute, a feat often mentioned in ancient Greek writings. They were dressed in gay costumes.

The Roman satirist Juvenal poked fun in his Fifth Satire at the African long-tailed monkeys which, dressed in green suits, rode on dogs' backs and danced for the amusement of the ladies. Martial, the 'scandalmonger' of the first century A.D., describes in his *Epigrams* the "wearisome habit of rich ladies" of dressing their tame monkeys in red- or white-hooded coats embroidered with gold braid, as though they were their own children. Similarly the philosopher Lucian (A.D. 125–180), who has been described as the "Greek Voltaire", made fun of the monkeys he saw dressed in costly fur-trimmed jackets, which were expected to play the flute or the lyre. It is interesting to note that this rhetorician and master of polished prose had to admit that the entertainment value of these superbly trained creatures far outshone that of many humans.

Plutarch mentions in one of his *Parallel Lives* that the Emperor Augustus never failed to be amused whenever he saw ladies going for walks with expensively attired pet monkeys in their arms or on a lead. Augustus is said to have asked one such lady, who looked as though she came from Greece or one of the regions in southern Italy inhabited by large numbers of Greeks, whether women in her homeland had given up having children, seeing that so many married women lavished all their affection on these creatures. Plutarch does not record the lady's reply.

In Book VIII of his *Geographika* the second-century geographer Claudius Ptolemaeus tells a similar story about the Numidian king Massinissa, who lived from 238 to 149 B.C. This nomadic prince, whose Tuareg squadrons fought alongside the Romans and were a decisive factor in the Second Punic war, allowed many Roman merchants into his country—now known as Algeria—after the final defeat of the Carthaginians. King Massinissa encouraged trade with Rome in every way and, in return for silver, supplied leopards and lions captured in the foothills of the Atlas Mountains, which in those days were thickly wooded, and also sold the Romans elephants and various kinds of monkeys. The demand for monkeys became so great that the King is claimed by Ptolemaeus to have inquired why it was that Roman women did not rear children of their own rather than monkeys.

Whether this ancient anecdote should properly be ascribed to the Numidian King or to the later Emperor Augustus is unimportant. What matters is that for centuries in Rome the demand for easily tamed monkeys must have been very great indeed, and that the interest shown earlier in the Greek states for these amusing creatures was maintained, perhaps even exceeded.

If we are to believe certain Greek and Roman historians, monkeys were exhibited standing in front of a board displaying cards, each with a letter of the alphabet painted on it. At a given word of command, they removed a number of the cards and arranged them to form words. If this is so, it must have been done by some clever form of deception on the part of the trainers, using tricks unknown to modern trainers.

Not one of the ancient writers mentions whether the 'spelling' monkeys were given triangular, circular, or square cards to use. If all three were used, then the monkeys' apparent ability 'at a given word of command' to select and assemble cards of a particular shape seems rather more plausible. That a monkey could have recognized letters of the alphabet in their true meaning is out of the question, as this presupposes a degree of perceptivity which not even the cleverest monkey possesses. It is more likely that the monkeys were trained, as were the 'spelling' and 'counting' horses of the twentieth century, to react in a certain way to various gestures and signs made by the trainer which were scarcely perceptible to the spectators.

At the extremely popular monkey-shows given throughout the Roman Empire, gaily dressed monkeys could, according to Pliny's *Natural History* (VIII, 215), be seen playing draughts. They sat at low tables on small stools, and as soon as their opponent—the trainer—made his move they in turn picked up a piece and placed it on another square on the board. The trainer explained every move of his opponents in a running commentary to the spectators, who were gullible enough to suppose that the monkeys were thinking out each move. But not all of them understood the game very well anyway and they were sitting too far away from the draught-boards to be able to assess the skill of the moves the monkeys made.

Several contemporary accounts of these games of draughts with monkeys were included by Konrad Gesner in his five-volume work *Allgemeines Thierbuch* (Universal Animal Book), which was published between 1551 and 1558. He mentions that "the monkeys also play chess" and quotes a number of sources, translated from Latin, to illustrate the cleverness of various species of monkeys.

In this connection it is worth mentioning that, according to a Jesuit priest named Harduin, the Emperor Charles V (1500–56) owned a monkey which had a personal servant employed solely to wait on it hand and foot. This monkey often played chess with the Emperor and is said on one occasion to have checkmated him. We cannot judge how accurate the Jesuit priest's account really is, but some indication may be gained from another assertion by this learned gentleman, that the monkey, having previously received a slap across the face from the Emperor, took a cushion next time it played chess with him and put it over its head for protection throughout the game!

Among the most amazing—and least credible—of the claims made by ancient and medieval writers regarding the high level of intelligence possessed by monkeys are those asserting the ability of certain long-tailed species and baboons to play musical instruments.

In ancient times the flute resembled the modern oboe. Flute-players usually played two flutes simultaneously. As the flutes were tied together just behind the mouth-pieces, the players' hands were free to play both at once. Judging by Egyptian, Assyrian, and Roman pictures and also by the information contained in Aelian's *On the Peculiarities of Animals* (V, 26 and VI, 10) and in other writers' descriptions, specially clever and imitative monkeys were trained to play such flutes. It is doubtful whether they could have managed to produce more than a succession of very unmusical noises, which in any case would have been lost amid the sound of the accompanying voices.

It would appear, from similar accounts, that monkeys were also trained to play the syrinx. This instrument, also known as pan-pipes or the shepherd's pipe, consists of from seven to nine reeds of varying length. It is conceivable that monkeys were taught to hold the syrinx and blow into it; but since they could not be taught to differentiate one note from another they could at best have produced only isolated sounds.

Similar results would have been produced with the lyre. This plucking instrument, not unlike the modern mandoline, had a large tortoise-shell case to increase the volume. The few sheep-gut strings were fastened between two 'horns' attached to the shell. The monkeys presumably held the lyre firm with one hand and plucked at random at the strings with the other. What is certain is that the monkeys must have derived just as much pleasure from the discordant sounds produced as the attentive spectators did from watching their performance. Such 'concerts' gave rise to the myth of 'musical' monkeys, and credulous historians reported such events without any deliberate intention to mislead their readers.

The Athenian comic poet Aristophanes (450–385 B.C.) describes in his comedy *The Acharnians* (960 *et seq.*) how at the harvest festival of Dicaeopolis camels were exhibited with tame monkeys riding on their backs. After performing their tricks the monkeys climbed down from their mounts and went round collecting the money thrown to them. Aelian (VI, 10) gives a similar description of a trained baboon. In an early fable ascribed to Aesop, dating from about 550 B.C., there is a reference to trained monkeys going round collecting money after a performance.

The monkeys which were exhibited in Rome and other parts of the Roman Empire were also trained to collect the copper coins known as *as* after their performances in the monkey theatres or in the open.

Pliny, Martial and Juvenal, as well as the Greek satirist Lucian, also mention the rewards collected from public performances by trained monkeys, sometimes carrying small shoulder-bags from which their owners later extracted the copper *as* and the occasional silver *sesterces* and gave their artists fruit and buns in return for their efforts.

In Juvenal's Fifth Satire (154 *et seq.*) there is an excellent first-hand account of a performance at a monkey theatre. The monkeys performed a variety of tricks, riding on dogs and goats and, protected by helmets and shields, throwing javelins at one another in the guise of Roman legionaries. The javelins they used were miniature reproductions with blunt ends, to avoid accidents not only to themselves but also to the audience when hurled wildly in the wrong direction. The spectators at these ancient monkey theatres must have had marvellous entertainment for their money, for, as Martial describes in his Fourteenth Epigram (202 *et seq.*), the monkeys would pick up the javelins thrown back by the spectators and hurl them straight back again.

Crude fun was much in vogue with the audiences of antiquity; nowadays of course it is frowned on by owners of performing monkeys. Lucian describes in one of his writings how once during a performance a young spectator threw a handful of nuts among the costumed monkeys. Their immediate reaction was to forget all about their trainer and their training. Only when they had eaten the last nut could the act be resumed. Such unscheduled interruptions reflect not so much on the monkeys as on the stupidity of those who cause them. Lucian declares that, so far as the spectators were concerned, the digression was immensely more entertaining than the set performance.

Sometimes monkeys were trained specifically to perform acts which poked fun at officialdom. The Roman moralist Apuleius described in his *Metamorphoses*, written about A.D. 150, how monkeys in costume appeared in a festive procession wearing masks resembling the faces of well-known high government officials. Furthermore, they wore the insignia of the dignitaries they were 'aping'—the consuls, praetors, and censors—such as the toga with the broad purple stripe worn by magistrates, while others carried lictors' *fasces* and the like.

It may well be that scientists in ancient times were aware of the existence of chimpanzees. In his *Natural History* (VIII, 216) Pliny gives an account of a kind of large African monkey, which he refers to as a 'satyr'. He describes it as being extremely docile, very manlike in form and with gentle human features. The Roman scholar made a distinction between this 'satyr' and the baboon and also the 'callithrix' (in Greek = 'beautiful-haired'), which may perhaps be identified with the Abyssinian guereza.

Monkeys kept as Pets in Ancient Times

Species	*Region*	*Acquired Skills*
Entellus	Assyria, Babylon, India.	Walking upright. Minor tricks. Said to have ridden on donkeys and dogs. Dancing.
Barbary Ape	Egypt, Carthage, Greece, Roman Empire.	Going round on a lead. Various tricks—riding on dogs, donkeys, goats. Dancing in costume. (The first animal to be subjected to vivisection: dissected by Galen.)
Green Monkey, Red Monkey, less commonly the Diana Monkey	Egypt, Carthage, Greece, Roman Empire (also possibly Asia Minor)	Performance in costume in the monkey theatres. Use of musical instruments. Riding on donkeys, dogs, among the Greeks also on camels, among the Romans on goats. Dancing.
Baboon (Hamadryas or Sacred Baboon. Less commonly the Dog-faced Baboon and Gelada)	Egypt, Roman Empire.	Going round on a lead. Riding on donkeys. Training in Egypt by temple priests. In Rome, a variety of tricks.

These animals were exhibited during the festival of the Consul Pompey in the Circus Maximus in 69 B.C. as 'wild men of the forests'. Pliny did not know what happened afterwards to these manlike creatures, but declared that nothing of the kind was ever seen again in Rome.

Pliny describes the 'satyrs', by which he probably meant chimpanzees, as "tame and capable of coherent instruction, but somewhat cunning". It is just these qualities which are more characteristic of chimpanzees than of baboons, long-tailed monkeys and Barbary apes.

CHAPTER THREE

WORKING FOR MAN

DURING the Middle Ages itinerant entertainers travelled around Asia and the Mediterranean countries with their trained monkeys. Various records dating from the eleventh to the fifteenth centuries show that they also made their way across central and western Europe. At the annual fairs at Augsburg and Frankfurt it was a common practice to exhibit tame monkeys as well as other animals in travelling menageries.

Over a thousand years separate the period when performing monkeys were to be seen in the Greek city-states riding on the backs of dogs, donkeys, and camels from the time when Charlemagne received a monkey, together with an elephant, as a gift from Caliph El Raschid in the year 801; and it was almost a thousand years later that long-tailed monkeys were exhibited in the seventeenth century at the court of the French King Louis XIV. Yet the tricks performed were in all probability much the same, and they were to delight the audiences of the monkey theatres in the eighteenth and nineteenth centuries no less.

In Europe the monkeys exhibited by travelling showmen at the fairs were applauded by admiring spectators for their cleverness and endearing ways. But in the newly discovered territories of America it had been the custom among the more primitive tribes from time immemorial to hunt monkeys—though not all species—for food. The Indian tribes which had made cultural progress, on the other hand, did not hunt them.

Long before the Spanish conquest of Central America tame monkeys were kept as house pets by the highly cultured Aztecs and Mayas, as indeed they were by some of the less civilized South American tribes. The species kept were the same as those hunted for food with blow-pipes or bows and arrows, still a common practice in the regions of the Orinoco and the Amazon.

Among the northernmost of the pre-Columbian civilized races, the Aztecs, and their neighbours, a monkey was the symbol of one of the days of the month. There is a parallel here with the ancient Egyptians, who represented the sacred baboon in their hieroglyphs on temple walls and in reliefs, just as the Aztecs represented in their pictorial script the white-shouldered monkey and the howling monkey. In their month of twenty days, the 'day of Ozomatl'—*i.e.*, the 'day of the monkey'—was the eleventh, dedicated to Xochopilli, god of flowers and mirth and patron of 'the merry monkey'.

Caravans belonging to 'pochtecatl'—armed merchants who penetrated far into the south to trade with the people of present-day Guatemala and Honduras—brought among other things tame monkeys from the primeval forests of Central America to the capital, Tenochtitlán, and presented them as gifts to eminent families whose favour they wished to gain. Monkeys were not among the animals eaten by the Aztecs.

Nowadays monkeys are very popular in South America as amusing house pets. For decades Central and South American species have also been exported to North America and Europe. Thousands of capuchins, spider monkeys and woolly monkeys are now owned by animal lovers in countries all over Europe. Just as suitable as pets are the highly adaptable squirrel monkeys and the small wistitis.

In modern times the earliest reports concerning anthropoid apes go back several hundred years. It is certain that long before the seventeenth century Europe knew about chimpanzees from accounts brought back by slave-traders, though at that time there was a total lack of authentic information about them. For the orang-utan, 'the red-haired man of the forests', from the

Dutch East Indies, a few vague, scattered details had been given by Cornelius Schouten, a Dutch captain, on his return from a South Seas voyage in 1616. (In the menagerie at the Czernin Palace on the Hradschin Hill in Prague a live orang-utan was being kept as early as 1582, but there is no detailed information about it.)

It was not until 1641 that the first live chimpanzee arrived in Europe. When it was exhibited near Leiden it caused a sensation. Its human behaviour and endearing ways were universally admired, and people flocked from near and far to see it. It would politely take any fruit or other food handed to it and eat it slowly and deliberately, but with obvious relish. Unluckily, after a few months it died of a lung infection, probably caused by the changeable weather and aggravated by being fed with the wrong kinds of food—bananas and other tropical fruits rich in vitamins had at that time hardly found their way into Europe. The anatomist Nikolaus Tulpius dissected the corpse, and the scientist Dr Olfert Dapper wrote a detailed description of it. (Over 130 years passed before another live chimpanzee was to be seen in the Netherlands. It came into the possession of the Prince of Orange, who installed it in his zoo, Het Loo, near the capital. In 1776 the first orang-utan to be shipped over from Borneo was brought to the same zoo. Consequently Dutch scientists like Vosmaern, who had himself owned a tame orang-utan in Borneo, were able for the first time to describe both species.)

Dr Dapper was an explorer as well as a scientist and he made several trips between 1650 and 1670 to different parts of Africa. His expeditions were financed by shareholders of the Netherlands East Indies Company, who were interested in obtaining first-hand reports on the possibilities of trading with the various African kingdoms, and in establishing trading stations and forts in well-situated harbours, to protect the sea route to Asia. In 1667 Dr Dapper published the first edition of his *Nauwkeurige Beschrijving der Afrikaensche Gewesten* (Curious Account of West Africa), in which he describes how apes were trained to work by the inhabitants of the West African coastal regions. According to his account, the natives captured young apes "which walk quite erect like men" and taught them to perform various tasks. They are said to have learned how to turn meat

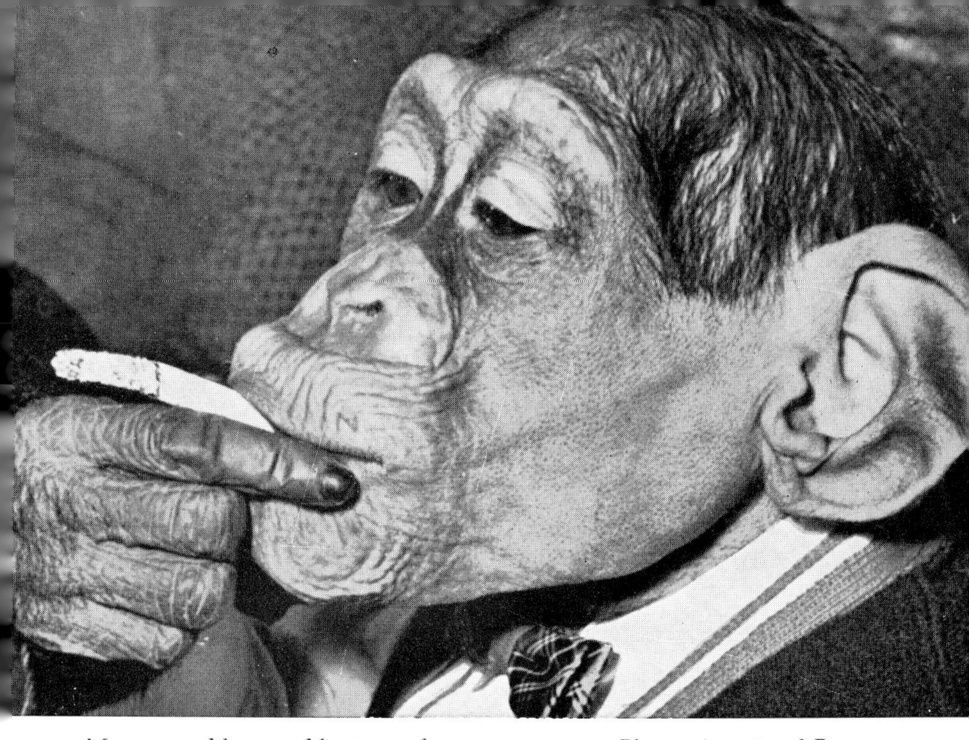

Many man-like apes like to smoke *Photo: Associated Press*

32

Three contented chimpanzees having lunch
Photo: Professor Dr B. Grzimek

A female chimpanzee feeding a wolf-cub *Photo: Professor Dr B. Grzimek*

33

Manpower is short, so Charlie stamps the mail *Photo: Ullstein-Seeger*

on a spit, grind millet in a mortar, and draw water in pitchers. The author goes so far as to assert that these apes would reproach themselves for being clumsy—for instance, if they spilled water from the pitchers—by uttering high-pitched screaming noises. It is true that he does not actually specify these creatures as chimpanzees, but they cannot have been any other species, since long-tailed monkeys, for example, do not possess the strength to carry pitchers.

The conscientious Dutch zoologist Jan Büttikofer also reported, in his book *Einiges über die Eingeborenen von Liberia* (Some Remarks concerning the Natives of Liberia) (1888), that he had seen chimpanzees there which had been trained to look after small children entrusted to their care by their parents: they not only played with the children, but also protected the rice-fields from unwelcome raids by monkeys—an easy task for them, since they regarded the rice-fields as their own domain and were therefore only too eager to drive off predatory intruders.

Diedrich Westermann, the author of many books on African territories, peoples, and languages, and former director of the African Institute of Berlin University, comments in a description of a West African negro tribe in Togo "that it kept long-tailed monkeys which were so well trained that they were used as salesmen". Alfred Lehmann, in his book *Animals as Artists*, quotes Professor Westermann: "Hollowed husks of gourds containing bundles of tobacco leaves each priced at 5 pfennigs are hung round the monkeys' necks and they are then sent off to the market; if a customer extracts a bundle from the gourd without handing over the money for it the monkey follows him around until he pays up."

Alfred Lehmann is in no way sceptical about this story, though he expresses surprise that the creatures used for the purpose should have been monkeys, rather than the more intelligent apes which inhabit the territory. Westermann is indisputably one of the leading authorities on Africa; Lehmann is equally an expert, in the training of apes and monkeys. Another animal trainer, Karl Faszini, was of the opinion that chimpanzees could easily have been taught such a practice but not long-tailed monkeys, save a few exceptionally talented individuals. Other trainers, from the period 1900–1920, have demonstrated that both Barbary apes and long-tailed monkeys, as well as macaques, can be trained to perform tolerably well the kinds of tricks expected of monkeys at circus shows.

As Alfred Brehm mentions in his *Illustriertes Tierleben* (2nd ed., 1876), the pig-tailed macaques of Sumatra are taught to gather coconuts. Brehm reproduces an eyewitness account sent to him by the explorer C. Bock, who spent several years in Borneo Celebes, Java, and Sumatra studying creatures in their natural habitats. Bock, in whom Brehm had every confidence, describes how one day in Sumatra he met a Malay coconut-gatherer with a tame pig-tailed macaque on a 16-foot lead. When he inquired whether the monkey had been trained to perform any special tricks the Malay replied that this was indeed so and that the monkey worked for him. Bock offered him a gift if he would allow him to watch the monkey at work. The owner agreed, and gave the monkey a command in Malay, whereupon it climbed up the nearest coconut-palm. On reaching the top it began to feel the nuts in expert fashion, picking only the ripe ones by twisting them off their stalks and throwing them down. It continued doing this until the owner told it to come down from the tree, which it did immediately. The man then gathered the nuts together and took them to his master. This was how the Malay, with the help of the macaque, earned his living. Bock emphasizes in his report that in Sumatra pig-tailed macaques are frequently used for this purpose.

Photographs and films of recent times have proved that Bock's account, written almost a century ago, is in no way exaggerated.

The oldest accounts of working monkeys originate from Egypt, and date back at least four thousand years. Frieze- and wall-paintings of that period show baboons gathering nuts from date-palms and doum-palms, and fruit from sycamore fig-trees. The fact that men are standing under the trees might suggest that the baboons are being used to harvest the fruit and nuts for them; it seems much more likely, however, that they are merely watching the baboons taking their pick for themselves. Even if the Egyptians knew nothing of vitamins they were undoubtedly aware of the nutritious value of freshly picked fruit, and would have preferred their very expensive and much treasured pets to have it rather than claim it themselves.

Another indication from ancient times, though of much later date, that apes performed tasks for their human masters is found in the writings of the Greek author Philostratus (*c.* A.D. 170–245). In one of his works he mentions the pleasure Indian monkeys derive from imitating, and how this was exploited to get them to pick pepper berries. We are not told whether the creatures in question were langurs or rhesus monkeys, but the description of what happened is interesting. First of all the Indian pepper farmers, when they knew monkeys to be in the neighbourhood, would pick a number of pepper berries and throw them in a heap on the ground. They would then go home, pretending not to see the inquisitive monkeys in the nearby trees. As soon as the farmers were out of sight the monkeys would climb down from the trees and imitate what they had seen done. The following morning the farmers would arrive to find all the berries gathered in a heap. Philostratus seems to have been unaware of the fact that these creatures naturally enjoy using their nimble fingers; he also wrongly assumed that they harvested the fruit during the night, seeing that Indian monkeys sleep at night. Even so, he may not, after all, have fallen for a tall-story, since recent observations of the same kind seem to support his account.

A detailed description of the pig-tailed monkeys living in Thailand and the Malaysian peninsula has been given in the magazine *Das Tier* by Helga M. Roth. These monkeys are skilfully trained to pick only fully ripe nuts and throw them down from the trees. Palm-trees in which they find only unripe nuts are quickly deserted in favour of others where the nuts are ready to pick. The account states that a single pig-tailed monkey can gather up to 800 coconuts in a day and so earn for itself quite a lot of money.

That macaque monkeys in Malaysia actually earn money for their labour is confirmed by 'Atticus' in the *Sunday Times* (November 13th, 1960).

John H. Corner, Professor of Botany at Cambridge, officially employed monkeys during his period of office as director of the Muncipal Gardens in Kuala Lumpur. Entered in the ledgers as "municipal employees", their job was to pick and throw down from the trees various rare botanical specimens. For each of these monkeys the owner received an annual salary of about 50 Malayan dollars, together with a free supply of rice, bananas, and hens' eggs. When the monkeys were not actively employed by the municipal administration they were hired out to managers of coconut plantations. Relations between the monkeys and their owners were very easy. Before starting work the monkeys were let off the lead, and after they had finished their tasks they returned of their own accord, allowing their leads to be replaced, just like well-trained dogs.

The use of chimpanzees to operate a conveyor-belt in a furniture factory in Texas was reported by the *Sunday Times* columnist George Schwartz on November 20th, 1960. According to the article, three chimpanzees were employed to stuff foam rubber into sofa cushions and zip them up. It was claimed that each of the chimpanzees did the work of three women. Objections to their being employed were raised not by any animal protection society but by the upholsterers' trade union,

Vital Statistics of Anthropoid Apes

Species	*Standing Height*	*Weight*	*Character*
Gorilla (West and Central Africa)	170–180 cm. (5 ft. 7 ins.– 5 ft. 11 ins.)	150–200 kg. (330–440 lb.) or more	Irascible and impulsive
Orang-utan (Borneo and Sumatra)	150–160 cm. (4 ft. 11 ins.– 5 ft. 3 ins.)	70–80 kg. (154–176 lb.) or more	Phlegmatic, deliberate and quiet
Chimpanzee (West and Central Africa)	120–130 cm. (3 ft. 11 ins.– 4 ft. 3 ins.)	55–75 kg. (121–165 lb.)	Self-confident, active and noisy

According to the *East Africa Year-book*, 1952, a mountain gorilla that had been shot was claimed to be 6 ft. 9 ins. tall (with an arm-span of 9 ft. 10 ins.) and to weigh 325 kg. (= 715 lb. or 51 stone).

which sent a delegation to the workshop to clarify the legal position regarding the use of animal labour in lieu of human labour. The article states that when the delegation entered, the chimpanzees, seeing strange faces, suspended work. The credibility of this report is impaired by the fact that there is no photographic evidence to support it. So long as there is a lack of documentary evidence to the contrary, the most that can be assumed is that this was an experiment to test the ability of chimpanzees to perform tasks of a strictly routine and repetitive nature. Serious-minded zoologists consider chimpanzees incapable of work conforming to set standards, however adept they may be at imitating, not least of all because of their sudden changes of mood.

CHAPTER FOUR

MEMBERS OF THE FAMILY

ONE of the first zoologists to make a personal study of manlike apes was the Frenchman Buffon (1707–88). While curator of the Jardin du Roi, later known as the Jardin des Plantes, in Paris, he kept a chimpanzee, which used a table-napkin to wipe its lips after eating or drinking. It would fetch a glass or cup and pour itself some wine without spilling a drop, though of course the bottle had to be already uncorked. It even clinked glasses with its human dinner-guests. At tea-time it took several lumps of sugar and placed them in the cup before pouring out the tea. Buffon describes the chimpanzee in considerable detail in one of the thirty-six volumes of his monumental *Histoire naturelle, générale et particulière,* emphasizing that it was very well behaved and always polite to visitors.

Cuvier, Buffon's successor as curator of the Jardin des Plantes and originator of the 'cataclysmic theory', according to which "the plan of Creation was gradually brought to realization by God", described chimpanzees and other apes in a very matter-of-fact way, whereas his opponent, Professor Étienne Geoffroy Saint-Hilaire, carried out a remarkable aptitude test with a tame orang-utan. Saint-Hilaire, a pioneer in the fields of anatomy and evolution, used the orang-utan to provide further evidence for his theory relating to the influence of environment on the changeability of species. First he let the ape watch as he tied and then untied a simple knot in a piece of string. This procedure caused the orang-utan a little difficulty to begin with, but it soon mastered it. Soon it became proficient in dealing with more complicated knots. It later came to regard it as great fun to untie a series of two or even three knots that had been tied separately in different places along the piece of string, loosening them at great speed. In his *Principes de Philosophie zoologique* (Paris, 1830) Saint-Hilaire gave a full account of this orang-utan's activities, stressing the fact that a human would have required more time learning to perform them.

Opening doors presented no difficulties, and it even learned in a very short while to open a locked door by turning the key, providing this was already in the lock. Saint-Hilaire then hit on the idea of removing the key from a locked door and hanging it with fourteen other keys on a ring. He then handed the orang-utan the whole bunch of keys. He was astonished to note how quickly the ape discovered by experiment which was the right key to fit that particular lock. Soon it was as adept at selecting the right key as he was himself, identifying it by its shape and size.

In another experiment a banana—an uncommon luxury in those days—was placed, in the presence of the orang-utan, under a stone so heavy that in spite of its considerable strength it was unable to move it to get at the prize. It was then shown how, with the help of a crowbar, the stone could be dislodged. After a few vain attempts the orang-utan succeeded in manipulating the crowbar properly, and was rewarded for its efforts with the banana.

The employment of a tool was declared by Saint-Hilaire to be purely a matter of imitation, not a conscious utilization, even though the ape was now able to repeat the activity whenever it chose to.

Orang-utans were just as uncommon in Europe during the whole of the nineteenth century as they had been in the days of Saint-Hilaire. Chimpanzees, on the other hand, were being imported in large numbers. Most numerous were long-tailed and rhesus monkeys. As early as 1380, Hans von Görlitz, the youngest son of the Emperor Charles IV whose court was in Prague, was the proud possessor of a trained vervet monkey. From the fourteenth to the nineteenth century vervets and related species belonging to the family Cercopithecidae, characterized by their long tails, were increasingly in evidence in Europe.

A showman from Augsburg named Trede, very famous in his day, gave exhibitions in 1838 at which numbers of long-tailed monkeys, rhesus monkeys, and dogs performed together, while in 1855 the monkey trainer Bruckmann presented an animal revue that caused a sensation because of its great variety of acts. Long-tailed monkeys rode on the backs of dogs, rhesus monkeys performed acrobatics on a trapeze, and ponies drew a coach decked out for a 'monkey wedding'. The highlight was a grand spectacle in which all those taking part in the revue captured a Lilliputian fortress by storm. The monkeys scaled the walls, while the dogs forced an entry across the drawbridge. The spectators were entranced as the final assault was made on the stronghold, and delighted when the assailants finally broke through into the main courtyard to claim their spoils—various sorts of food placed there in readiness for them.

Even before the publication in 1870 of Brehm's *Tierleben* there were several descriptive books on animals, all of which contained a wealth of information about monkeys, most of it inaccurate. Darwin's Theory of Evolution was at that time being interpreted in a very arbitrary way. Nevertheless, primates were now being designated as quadrumanous (four-handed) and no longer as bimanous (two-handed) creatures. Arguments were raging about the concepts of "transmuted humans" and "man as the perfectly developed cousin of the ape" (as Brehm assumed). Besides the question of the relationship existing between man and ape there was also contention as to whether certain apes, such as the baboon, used stones as missiles and whether they were at all capable of using sticks or stones as tools. Brehm denied them this capacity, but conceded that many Europeans resident in Africa had claimed to have confirmed this from first-hand observation. His cautious assessment of such reports was due to the fact that he had from personal experience formed strong reservations concerning the eyewitnesses in question, affected as they were by the climate and their excessive devotion to the whisky bottle. In spite of this he reproduced in his work a number of these accounts, some of which were later called in question by editors of successive editions. One account, regarded as spurious, was of an extremely intelligent female chimpanzee said to have been trained to perform various tasks on board ship, including heaving up the anchor and reefing and furling sails; she not only kept the fire in the baker's oven stoked up but also gesticulated to the baker when the required temperature was reached!

This account seemed so improbable to Professor Pechuel-Loesche, who in 1893 brought out a third, completely revised edition of Brehm's *Tierleben*, that he relegated it to a footnote simply stating: "Degrandpret saw on board a vessel a female chimpanzee which was extraordinarily intelligent and well trained and performed a variety of tasks."

Another account included in the first edition but rejected in the third was an eyewitness report by a man named Brosse, who claimed that during the return voyage from Africa to Europe a pair of chimpanzees ate their food using knife, fork, and spoon just as human beings do, and drank all kinds of beverages, including wine and brandy. The Professor considered as even more unlikely the claim that the chimpanzees regarded the cabin-boys as mere servants and let them know by "gestures" whenever they required anything.

Professor Pechuel-Loesche was himself a zoologist of some standing, having taken part in the 1874 Luanda expedition to observe chimpanzees in their natural habitat and seen for himself a number of aspects of their communal life, which had surprised him greatly. Yet he suppressed both these stories. He was not to know that fifteen years later trained chimpanzees in circuses would be demonstrating just how proficient they really can be with knives and forks.

The Professor nevertheless accepted Brehm's assertion that one cannot treat such creatures as mere animals but rather as higher forms of life influenced by the behaviour of the human beings with whom they have frequent contact. Similarly, Brehm's contention that a chimpanzee understands what is said to it was accepted by him, as was the assertion that in its relationship with man the chimpanzee submits to his superior natural gifts and abilities whereas in its relationship with other animals it adopts the same attitude as man does towards it, considering itself on a higher plane, especially if the others belong to a related species.

As for his own chimpanzee, Brehm reported that when it was in a happy frame of mind it could be observed quite clearly to grin, though it never actually laughed, something Buffon claimed to have observed. Modern trainers agree that less highly developed species than the chimpanzee show their pleasure by facial expressions, but chimpanzees are able to display cheerfulness or reflectiveness, or even displeasure.

The chimpanzee trainer Karl Faszini, who had almost twenty years' experience training apes and living among them, observed that chimpanzees behave differently when they feel they are being watched from when they believe themselves to be alone.

During their stay in Sydney, Faszini and his wife noticed one day, on returning home after taking Charley the chimp for a walk, that some bananas and other fruit were missing from a locked chest. The disappearance of the fruit was something of a mystery, as the key to the chest was always kept in Faszini's jacket pocket. Whenever he changed into a suit to go out he hung the jacket on a hook near the door. Faszini explains the mystery as follows:

> Charley and his companion Susi occupied the next room. Anyone who does not know these chimps would no doubt find it difficult to understand how we could leave them both at home without any supervision for hours at a time. But we had become so accustomed to their behaving themselves that there appeared to be no reason why they should not be left on their own. A few days after I had discovered that the fruit was missing my wife went out again with Charley. I remained at home, and was just settling down to relax when I heard faint noises coming from my living-room. I got up, crept over to the door, and peered through the keyhole. I saw Susi in the act of turning the key in the lock of the chest. She raised the lid, took out a number of oranges and bananas and placed them on the floor. Then she closed the lid down, extracted the key, and replaced it in my jacket pocket. Having done this she went into her own room with the fruit. When I opened the door and walked into her room she was just starting to peel a banana. She looked up in surprise. I spoke a few words to her in a friendly voice, and feeling reassured she proceeded to

eat the banana. I continued to leave the key in my jacket pocket. There were never more than a few bananas or oranges missing—Susi remained reasonably modest in her requirements.... She only forgot one thing—to lock the chest after her!

Another story told by Faszini shows how inquisitive chimpanzees are. He and his wife left one of them alone in a room, to see how it would react when it thought it was not being watched by anyone. They locked the door and walked away making lots of noise to show that they were not spying. They saw the chimpanzee watch them intently as they walked along the pavement past the house and out of sight round the corner. They then crept back into the house in stockinged feet, and Faszini peered through the keyhole. He was highly amused to find himself looking straight into the eye of the chimpanzee! Despite their efforts to make the least possible noise it had evidently heard the couple return and had placed its eye at the keyhole as though knowing that its master would do the same.

Although in the nineteenth century chimpanzees were a fairly common sight in most European countries, people's knowledge of gorillas was based largely on the highly embellished stories recounted by European big-game hunters returning from darkest Africa. It was not until the last quarter of the nineteenth century that authentic material about gorillas became generally available. This was mainly due to the arrival in Europe of the first live gorilla, which natives accompanying it on the voyage from Luanda had christened Mpungu. It was a present from a group of Portuguese planters from Pontanegro to the German doctor and zoologist Falkenstein as a token of their gratitude for the way in which he had treated a number of their compatriots who had fallen desperately ill. The doctor had accompanied the German Luanda expedition of 1874–75 into the frontier region north of the mouth of the Congo between the French and the small Portuguese possessions, with the object of investigating the animal and plant life of the region.

Soon after Mpungu, whose mother had been killed by native hunters, had been handed over it became clear that he was not only a lovable pet in need of affection, but also, in spite of being still very young, highly talented. He looked on the doctor as his protector and provider and obeyed him without hesitation. Allowed to roam freely round the camp, Mpungu was fed on goat's milk and fresh fruit and gradually became accustomed to the same food as the doctor ate. From him he learned how to drink from a glass or a cup and eat from a plate, and generally displayed the same good manners at table as are sometimes observed with tame chimpanzees. He showed his pleasure by romping about and when in a good mood would often beat his chest with the palm of his hands. His favourite playmates, apart from a number of natives and Portuguese who had befriended him and, of course, the doctor, were some chimpanzees and the goat which provided his milk.

After nine months getting used to living with humans Mpungu was sent across to Europe by Dr Falkenstein in June 1876, arriving in Berlin while the weather there was favourable to acclimatizing him to his new surroundings. Two well-heated rooms had been provided in advance for him by Dr Hermes, director of the Aquarium, and visitors were able to watch him through a thick glass partition.

The same kind of accommodation as was provided for Mpungu is still used in the monkey houses of zoos; the glass partitions are intended not so much to afford protection to visitors as to the valuable gorillas, orang-utans, and chimpanzees, which are very susceptible to people's germs. It also provides a safeguard against those enthusiastic but misguided visitors who feed the animals, often with titbits which do them far more harm than good.

In those days, however, the fundamental principle of segregation from the public was certainly not observed. Mpungu was allowed to play with the keeper's children and was even carried about in the arms of visitors. At this period Mpungu weighed about 40 lb. and was 28 ins. in height. Stretched out at full length he measured 32 ins. from top to toe. He was also allowed to pay occasional visits to the big monkey-cage. A lecture by Dr Hermes delivered to a meeting of German naturalists and doctors, included the following comment:

> In the communal cage he is the absolute lord and master, and even the chimpanzee submits to his superiority without

any show of resistance. Mpungu treats the latter on equal terms, but scorns the rest of the community as just rabble. . . . In his day-to-day life Mpungu takes his cue from his keeper, and has the same meals as he does. For breakfast he has a couple of sausages or a cheese sandwich, washed down with a glass of good white wine. It is fascinating to see him grip the large glass with his short, stumpy fingers, and indeed it would fall out of his hands if he did not use one of his feet to help him hold it steady. At about one o'clock the keeper's wife brings him his lunch. Throughout the hot summer, while he was living at home with me, he could hardly wait for one o'clock to strike. If the door-bell rang he insisted on opening the door himself.

For lunch Mpungu began with a cup of meat broth and followed this with the main course, consisting of meat and vegetables, for which he used a spoon. He would follow this up either with a piece of roast chicken, of which he was especially fond, or with fruit, which he liked equally well. During the afternoon he was given more fresh fruit, and in the evening some slices of bread and butter, sometimes also a few sandwiches, with either milk or sweet tea. In the light of what is now known of gorillas in captivity, this diet was totally wrong.

His active day lasted from 8 a.m. till 9 p.m. He had a soft mattress to lie on and a brightly coloured woollen blanket, in which he carefully wrapped himself, even for the 1½ hours' siesta after lunch. "The keeper sits beside him until he is asleep, which, in view of his great need for sleep, is usually not long. Best of all he likes sleeping in a bed with his keeper, whom he embraces; he snuggles down with his head on the keeper's chest."

When Mpungu fell ill, however, he was much less affectionate and struck out at his keeper and bit the doctor treating him. He had an aversion to quinine on account of its bitter taste, and would only drink it if it was sweetened with sugar and mixed with mineral water. Naturally, Mpungu was not treated by a vet but by a children's doctor. On one occasion he was even visited by Professor Virchow, the most celebrated pediatrician of his day. When he died he was only sixteen months old. He had cost 20,000 marks.

One of the most memorable events in his short life was a

tragicomic encounter he once had with a scientist. For several years the zoologist Johann von Fischer had been making a study of the sounds uttered by baboons and rhesus monkeys. From his accumulated notes and observations he believed that he had reached the stage where he could converse with any species of African or Asiatic ape or monkey, as it was his considered opinion that there is only one basic monkey language, shared by all species.

When he heard of Mpungu, he contacted Dr Hermes with a view to trying out on the young gorilla the vocal sounds he had learned from the baboons and rhesus monkeys. As Dr Hermes was himself interested in such an experiment, permission was granted, and at nine o'clock one morning in 1876 Dr Fischer was introduced to Mpungu in the Berlin Aquarium. He reported on his attempt to converse in the supposedly universal monkey language:

> I sat down next to him on the bed, took him on my lap, and proceeded to apply my whole range of monkey language in order to induce him to participate and also to study his facial expressions. But in vain; whereas a baboon or a macaque would have understood me immediately and replied in kind, he merely looked at me in amazement, seeming to discern in my efforts nothing but a meaningless babble accompanied by grotesque grimaces. He himself did not utter a single sound, but at length grinned and gave me a cuff on the ear, followed immediately by a bite on the nose.

It is very likely that Mpungu would have understood a good deal more if he had been addressed in the human tongue familiar to him. It is also possible that the sounds and grimaces, perhaps intelligible to a baboon, had an unfriendly significance to him which he found infuriating.

As a result of his abortive experiment, Dr Fischer became the butt of a good deal of ridicule, and he was advised to perfect his knowledge of the gorilla language before embarking on any further conversations with Mpungu. A quarter of a century later the animal writer Dr Zell, in his book *Are Animals Unreasonable?* (1903), drew the false conclusion that a gorilla is unable to understand 'monkey language' because it does not belong to those species which live communal lives.

Today we know that all species of apes and monkeys have their own individual ranges of sounds and accompanying gesticulations. Those of a baboon are as different from those of a chimpanzee as they are from those of, say, a macaque. The vocal utterances of baboons or macaques would be regarded by a gorilla as 'foreign languages'. In Dr Fischer's day what was known about gorillas was still largely erroneous. For example, it was seriously believed that they abducted women, a widely accepted myth which figured in the work of such reputable artists as the Frenchman Emmanuel Frémiet and the German sculptor Gustav Eberlein. It has since been proved beyond all doubt that such stories belong purely to the realm of fiction, though the legend lives on in films and a certain genre of literary hack-work.

Over the past few decades a large number of gorillas have been kept in zoos and a careful check maintained of their age. One gorilla named Bamboo died in Philadelphia Zoo at the relatively advanced age for the species of about thirty-five.

The New York Zoological Society has in recent years broadened the scope of the Bronx Zoo (which alone has three million visitors a year) and the Aquarium to provide facilities for closer scientific observation. It is some years now since it set out to enlighten Americans on the true character of the gorilla, as well as that of the chimpanzee and orang-utan. The life and behaviour of these three species have been extensively portrayed in popular scientific films and television programmes. As a result, the gorilla has come to be seen in an entirely new light. From all the evidence it becomes clear that to call a gangster a 'gorilla' is to do the gorilla an injustice. However wild-looking it may appear, it is in fact an extremely peace-loving creature which as a rule shows great affection towards its keepers and other people who have gained its confidence. It is only when it senses that its family life is being threatened in some way that it is likely to turn nasty, a quite natural reaction shared by all the primates and by man himself. Its keepers, however, need to bear in mind at all times how enormously strong it is when fully grown, and to remember that it can break a man's spine or crush his ribs when giving vent to anger or just through sheer exuberance.

At Lincoln Park Zoo, one of the two zoos in Chicago, a large gorilla named Bushman was especially fond of one of his keepers. He was, as Vance Packard explains in his book *The Human Side of Animals* (1951), beside himself with pleasure whenever this particular keeper spent time with him. Bushman would greet him with cries of joy even if he was out of sight and he could only hear his voice. But the keeper never again ventured into the cage following an incident when Bushman clasped him in his powerful arms to show his affection—an embrace which made his ribs ache for a long time afterwards. Bushman showed unmistakable signs of jealousy, by various gestures and utterances, whenever 'his' keeper paid any attention to other apes. He died at the age of 23 in 1951, weighing 39 stone.

The New York Zoological Society has also given an account of a five-year-old female gorilla which spent its early years with an American family, enjoying equal rights with all the human members, so that it adopted many of the habits of civilized human society. Just as the orang-utans and chimpanzees which live in close contact with humans learn by copying, so did this young gorilla; soon it knew how to take a bath or a shower, and how to flush the toilet. It drank milk, tea, and soup from a cup and used a spoon when eating. It was meticulously tidy, always throwing rubbish in the dustbin. In this respect it resembled the orang-utans which were reared with the family of a planter in Borneo and were later transported to Hagenbeck's animal park near Hamburg. These facts prove that the higher species of apes are capable, within limits, of being 'civilized' when young.

It is not only chimpanzees living in constant contact with humans that show signs of civilized behaviour. In one of the American monkey stations a completely tame and highly 'civilized' chimpanzee named Jojo always switched the light off itself before settling down to sleep; Karl Faszini's chimpanzees did the same; and so did this young female gorilla. After leaving the bathroom and before getting into bed, in a room of its own, it always first switched the light off.

Certain intelligence tests have been tried out at the Yerkes Laboratories of Primate Biology at Orange Park on orang-utans and chimpanzees, and also on gorillas. Dr Robert Yerkes

Photo: Conti Press

Kasper the chimp heard that a strong man was wanted to stir the cake-mix

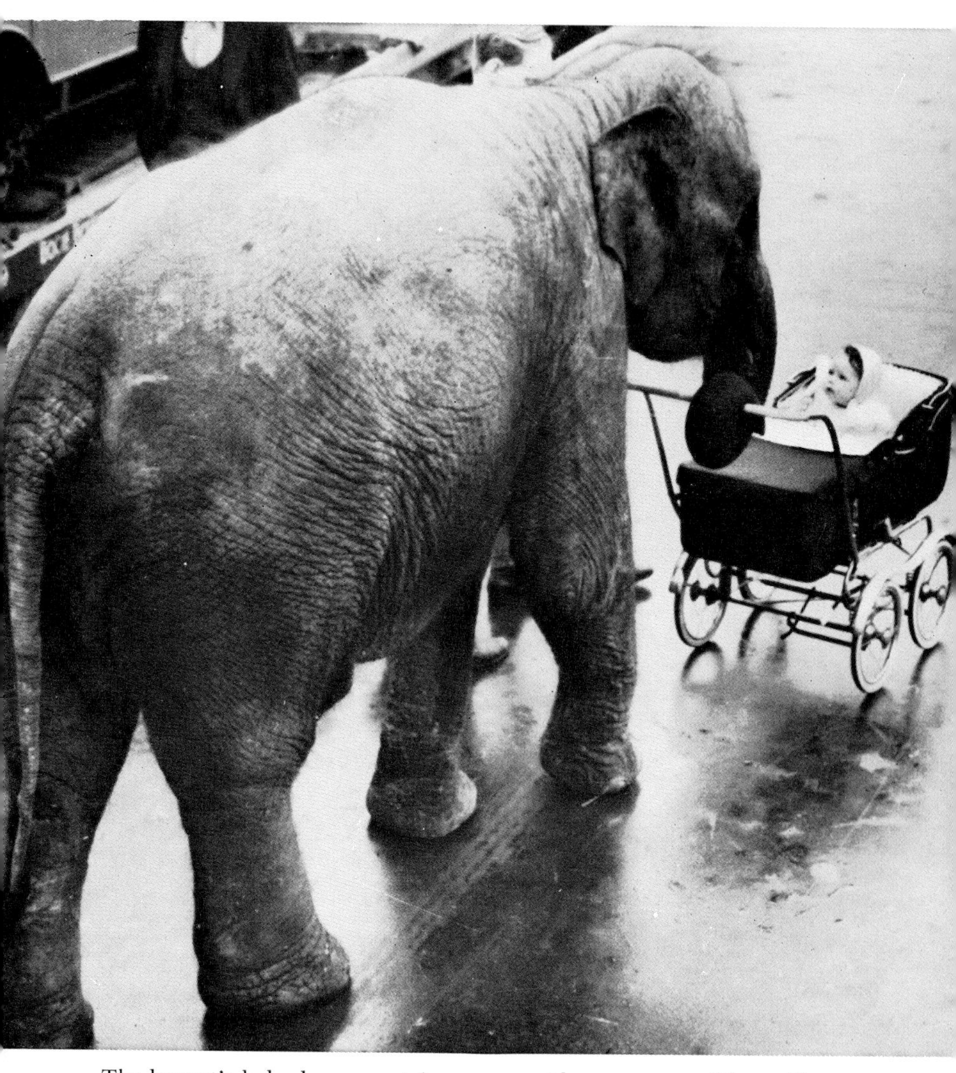

The keeper's baby has an outsize nursemaid *Photo: Keystone*

was particularly successful in a series of tests with a gorilla named Congo which indicated that in some respects gorillas are closer to humans than chimpanzees are. The same basic conclusions have been reached by a number of German zoo directors and also by Dr Ernst Lang, director of Basle Zoo, who has stated: "That the gorilla is our nearest relative in the animal kingdom cannot be disputed."

Even though chimpanzees may be the easiest of the primates to train to perform tricks, they are surpassed by gorillas when it comes to tidiness. For instance, Congo always unwrapped sweets in exactly the same way as an orang-utan or chimpanzee would, but whereas they would all just drop the wrapper Congo would screw the paper up and carefully put it with the other rubbish. Like Pongo, the gorilla killed near the end of the Second World War in Berlin Zoo, and his predecessor Bobby, Congo was a stickler for tidiness.

When it was raining and Congo was taken from his cage to the open one where many of the tests were carried out, he would return to his quarters to collect an armful of straw. He would spread this out on the wet floor, and only then would he be prepared to sit down.

Congo was given a test commonly tried out on chimpanzees. Boxes are provided which can be piled one on top of another to reach a banana. He took more time to stack them, but once he had done so they never once fell down, whereas the boxes stacked by the chimpanzees frequently collapsed.

The first gorilla ever to be born in a zoo was born in 1956 in Columbus, Ohio, and the first gorilla born in a European zoo was Goma, who was born at Basle Zoo in September 1959. At the time his mother, Achilla, had been resident there for eleven years. She had become headline news in 1952 when she had a stomach operation for the removal of a pen she had swallowed; the operation had been carried out exactly as for a human, including anaesthetics.

Dr Lang observed that during the first two years of Goma's life he was interested only in children, whose behaviour he seemed to copy closely. For his first birthday a number of boys and girls who had become friends of his were invited to a party at the zoo. Between them they ate a large cake, but not before

Goma had eaten a bouquet of flowers which had been presented to him as a birthday present.

Dr Lang wrote a book entitled *Goma, the Gorilla Child,* in which he makes many interesting comments about Goma and the two other young gorillas which were later to keep him company. Photographs taken by Dr Lang himself confirm that Goma at the age of two could perform a number of things which a human youngster cannot until about the age of four. Dr Lang remarks: "Goma will always look on human beings as being of his own kind."

On April 17th, 1960, Goma's mother Achilla gave birth to her second offspring, another male, which she herself cared for from the day of his birth. By the time Jambo, as he was christened, was a few months old his mother had taught him quite a few games, which they played together rather boisterously. When their keeper or Dr Lang entered the cage Achilla would thrust Jambo into their arms and watch rapturously as they played with him. Dr Lang also frequently observed Achilla hold up Jambo with motherly pride to the chimpanzees in the next cage.

Even if, so far, gorillas have not been tamed and trained to the point where they could safely be presented in circus acts and the like, and although it is true to say that the only gorillas known to have shown any real aptitudes have been those reared from an early age by African farmers or in zoos, it is by no means improbable that at some time in the future a gorilla born in a zoo will show that it is capable of performing at least as well as orang-utans or even chimpanzees. For the time being, however, only these two species can display their talents in public.

CHAPTER FIVE

LIVING IN SOCIETY

DURING the 1890's, for the first time in an American circus, a chimpanzee was presented as the driver of a 'Roman chariot'.

Shortly after the turn of the century, Consul, a chimpanzee which had been trained by an Englishman, performed a number of remarkable tricks. The most important chimpanzee trainers in the years that followed were Karl Faszini and Ernst Perzina, who owned a troop consisting of a female orang-utan and several chimpanzees. In both troops chimpanzees were exhibited which could ride bicycles, perform acrobatics, and have tea-parties.

Faszini made numerous observations based on his acquaintance with primates acquired at an early age, whose confidence he had gained in a way normally possible only for zoo keepers. Anyone who, like Faszini, has lived in the closest possible contact with trained apes over a period of twenty years is apt to become so accustomed to their various idiosyncrasies that he tends to take for granted certain habits which would be of absorbing interest to students of animal behaviour.

During the years 1904–24 Faszini specialized in training chimpanzees. His most successful 'pupil' was a dark-haired male named Charly, 4 ft. 3 ins. in height and weighing about 9½ stone. For a human of comparable height Charly's weight would have been somewhat excessive. He lived among other members of the same species for about eight years, but far outshone them in intelligence. He had excellent table manners, eating with a knife and fork as well as with a spoon. Like the other chimpanzees in the troop Charly slept on his own mattress and had a blanket, a tartan linen coverlet, and brightly coloured check sheets.

Faszini's chimpanzees used handkerchiefs and dusters, just like Brehm's tame chimpanzee, but they also used hand towels after meals or after washing their hands and face, unlike Brehm's pet, which used a handful of hay for the purpose.

They also enjoyed smoking cigarettes, which Faszini would light for them first. In time, however, they learned to light cigarettes themselves from a lighter held out to them. They always placed their cigarette ends in an ash-tray, even learning to put them out themselves. (A number of chimpanzees placed in Dr Grzimek's care in Berlin during the Second World War also learned to smoke cigarettes.) Thus, whenever a chimpanzee trainer has to submit his tax forms he always mentions—quite justifiably—cigarettes as well as the usual items such as food and drink (including moderate quantities of beer and wine). On more than one occasion Faszini has had to prove to sceptical inland revenue officials that his chimpanzees were compulsive smokers.

The chimpanzee which Cherry Kearton acquired during an African expedition soon learned to become a pipe-smoker, consuming half an ounce of tobacco a week—equivalent to 14–15 cigarettes.

At St Louis Zoo in 1950 the training of chimpanzees turned out to be so successful that one of the troop regularly served the others at table like an experienced waiter, bringing the tray laden with cups and saucers to the table, placing the food before each of the diners, and carrying out a number of other tasks before joining them. When the keeper wanted to light a cigarette he only had to hand him a box of matches and he would take out a match, light it, and offer it to the keeper. In return, he expected to be given a cigarette himself. An orang-utan in the same zoo enjoyed smoking cigars, though in this case the keeper had to light them for him.

Some of the chimpanzees at the St Louis Zoo used to ride around in a car which they drove themselves in a roped-off part of the grounds. Before leaving their quarters for a spin they would put on special clothing. Their observance of the Highway Code was every bit as good as that of many human motorists, though it is true that they enjoyed crashing deliberately because of the noise it made; however, little damage was done, for their vehicle was very solid and could only go slowly on its low-power electric motors.

The highly intelligent chimpanzee Duke lived in his keeper's house, helped himself to food from the refrigerator, made his own bed every evening, performed his ablutions with maximum decorum, and switched the light off before going to sleep.

For sixteen years the clever and highly self-confident female chimpanzee Missie lived at Berlin Zoo. Dr Ludwig Heck described her as almost human, for example in the way she fetched her own cups and plates from the cupboard, which she unlocked herself. In his book *Bobby the Chimpanzee and Other Friends of Mine* (1931) he writes:

> Anyone watching our Cameroons chimpanzee Missie sitting at table in her salon, pouring out three cups of coffee one after the other, and then smoking a cigarette, having lighted it herself, must have had an uncontrollable urge to laugh, as her antics were irresistibly comic. But she also gave food for thought to scientifically-minded people.

Dr Heck was extremely fond of trained animals and never missed an opportunity to attend a performance at the Wintergarten, at that time the largest music-hall in Europe, when they they were on the programme. For the same reason he would go to the circus at least once a month. He had made the acquaintance of the trained rhesus monkeys belonging to Ernst Perzina's troop, "which at times became real crosspatches" but which were highly skilled at performing gymnastic feats on the horizontal bars and the trapeze. Later, in 1910, he again met Perzina, who by this time was exhibiting an orang-utan and a chimpanzee riding a tandem, the chimp seated behind the driver in front and Grete, the orang-utan, installed in the rear saddle, both dressed in Tirolean costume. Every performance was a sell-out, and the antics of the chimp and Grete were for several weeks the talk of Berlin.

Dr Heck was at that time of the opinion that these performances were overrated; for "we know from the zoos how easily chimpanzees and orang-utans learn. Not only in their physical features but also mentally they are the animals closest to man." He also recalled how the chimpanzee Consul and his trainer had taken a comfortable room with bathroom attached at the Hotel Monopol, and that there, amid all the most up-to-date amenities devised for human comfort, Consul had succumbed to a lung infection. Doctors had treated him exactly as if he were a human being, and he had swallowed all the medicines prescribed for him like a well-behaved child, and yet he could not be saved.

Other chimpanzees in various American primate experimental centres also behaved well during a series of medical examinations. The cleverest male at the Yerkes Laboratories of Primate Biology at Orange Park was called Moos. He was acquainted with all the doctors and veterinary surgeons at the Laboratories and regarded them as helpful friends whenever he had a pain. One day Moos repeatedly pointed to his mouth until finally his keeper informed a vet. When he arrived, Moos opened his mouth wide, seized the right hand of the vet and guided his index finger carefully to the spot where a molar had set up a swelling as it was just breaking through the gum. Moos was duly treated.

Vets in zoos sometimes get their first inkling that an ape needs treatment by watching its behaviour and that of the other inmates of the same age. One vet at St Louis Zoo noticed how one chimpanzee was trying in vain to extract a loose tooth from another; when the vet entered the cage and went over to help, the chimpanzee allowed its tooth to be extracted by him without any fuss whatever.

Yet not all chimpanzees prove to be willing patients. If their first experience is painful they try to avoid further medical treatment. Dr Henderson, at one time the vet for the Ringling Circus, once found himself obliged to inject the very popular chimpanzee Nelly with an antibiotic to save her from certain death through a lung infection. Dogs can usually be held down without much difficulty, and other animals can be given a dose of chloral hydrate in a sweet, milky solution to send them off to sleep. I myself saw this happen in 1950 with a bear belonging to the Bügler Circus in Essen. It was give 24 grammes of the drug—enough to put three horses to sleep—after which the injection could safely be administered. Unfortunately chimpanzees, usually such trusting creatures, often simply refuse to take medicine, which they view with deep suspicion. In order, therefore, to give Nelly the treatment she so urgently needed, Dr Henderson asked two of the keepers whom she knew very well to take her by the hand, one on each side, and lead her out of the surgery. At a given signal a third keeper grabbed her feet from behind, while the other two struggled to keep her arms outstretched to avoid being bitten. The illness had weakened Nelly considerably, so that it was not too difficult for Dr Henderson to give her the injection in the buttocks. This rather rough handling produced the desired effect, for Nelly was soon well again. But never again did she allow both her hands to be held at the same time. Her painful experience taught her always to keep one hand free.

In the 1920's Cherry Kearton reported many instances of the clever behaviour of his chimpanzee Toto. He claimed that of the innumerable chimpanzees he had known Toto was by far the most gifted. By watching his master he learned how to use a hairbrush, and often washed his face, hands, and feet (though he was less keen on washing himself all over), and quickly learned to drink from a beaker. During their long treks through Africa he helped his master and the natives by carrying small loads on his back.

By breaking off the twigs and leaves from a branch Toto made a stick for himself, which he used to kill rats. He also made use of sticks and stones to ward off dogs.

Toto had often seen his master holding out his right hand to greet friends. He in turn cultivated the habit of extending his own hand as soon as his master's greeting was completed, though he sometimes became confused as to which was his right and which his left hand.

Kearton describes how Toto showed himself on one occasion to be particularly clever, in the way he extracted a cherry from the bottom of an empty wine bottle. He took a long, straight chicken bone, poked it into the neck of the bottle which he then turned upside down, and slowly coaxed the cherry out.

When Kearton was ill and obliged to stay in bed, Toto would obey his master's requests for a glass of water or the flask of quinine. In *My Friend Toto* (1931) Kearton also describes an occasion when Toto untied his bootlaces, removed his boots and placed them on the floor. He could also fetch a particular book from a whole row on a bookshelf. This was done by Toto placing a finger on each book in turn until Kearton indicated the right one by saying "Yes". According to Kearton, Toto understood many words and even whole sentences in English, French, and Swahili.

There is plenty of evidence that primates can recognize pictures. Miss Lilo Hess, the animal photographer, had a young female chimpanzee named Petra who could distinguish quite clearly in illustrated magazines and books between pictures of apes and other photographs and drawings. As Petra had a kitten for playmate, she was also able to recognize pictures of cats, "even when they were dressed up as they often are in storybooks", as Miss Hess explains in her book *Petra, My Chimpanzee Child*. When she saw her own reflection in the mirror she reacted just as all apes and monkeys do, by first trying to touch the 'other monkey' in the mirror and then going behind the mirror to try and locate it there.

Soon after the animal park was opened in May 1907 at Stellingen, near Hamburg, its founder, Carl Hagenbeck, succeeded in acquiring two fully tame orang-utans from a Dutch planter in Borneo. The planter had reared them on milk, and for seven years they were brought up as accepted members of the family, so that they became entirely accustomed to being in the company of human beings. In his book *Beasts and Men* (1909) Hagenbeck comments:

> At midday they dined at table with their master and received the same food as the family—in short, they were treated just as if they were human children, and they behaved themselves well at table. On their way over to Europe they were allowed to roam freely on the boat at night, and it was not long before they were the darlings of the whole ship's company.

Soon after arriving at the animal park, Jakob and Rosa were introduced to a chimpanzee named Moritz, who was about the same age. Moritz got on famously with both of them, soon proving himself the liveliest of the trio.

With the help of Rosa, Moritz succeeded on several occasions in scaling the high, smooth wooden fence and entering the adjoining giraffe enclosure. Working together, Moritz and Rosa would lift a heavy metal ball on to the solid wooden box which served as a bed. Then, while Rosa stood on the ball, Moritz would climb on to her shoulders, making it possible to reach the top of the fence. When the height of the fence was raised the chimp made use of a climbing-rope which hung from the middle of the cage ceiling. Clinging on to the rope, he made it swing to and fro till, with split-second timing, he released his hold on it and went soaring to the top of the fence. The fence was then extended right up to the ceiling. Moritz lost no time finding another ruse to enable him to continue his unauthorized excursions to the giraffe enclosure: he filched the keeper's bunch of keys and tried out one key after another until he discovered the one which fitted the lock. The keeper, knowing Moritz to be not only very clever but also quite harmless, stood by smiling while Moritz experimented with his keys.

Jakob, too, showed occasional flashes of intelligence. Hagenbeck has reported the notable occasion when Jakob used an iron bar as a lever to force the handle of a lock. The trio had broken off a length of iron during one of their gymnastic displays and subsequently used it quite deliberately as a tool, using their combined strength to break the lock as described above and then escaping into the park grounds through the open cage door.

This incident was written up in 1909, long before the intelligence tests carried out by Dr Wolfgang Köhler in the monkey stations on Tenerife were ever thought of. Similarly, Hagenbeck's descriptions of the table manners of primates were published long before Dr Köhler began his series of experiments. Sometimes Jakob and Rosa were even served at table by Moritz. He would bring a tray laden with plates of food, which he would serve. When the meal was finished he would carefully collect the plates and hand them to the keeper to take away.

> The trio receive for their lunch exactly the same food as I myself have at home. They like their food, and have grown used to good plain fare, which they find extremely appetizing. Sometimes they have a good red wine, watered down, with their meal. Jakob has shown himself to be the one fondest of

wine, while Rosa has shown a certain ladylike disdain for alcohol.

Hagenbeck often observed how Rosa, Jakob, and Moritz collaborated to tease the spectators:

> The two orangs, who prefer to sit by the bars, have a habit of stretching out their hands to greet visitors. As soon as a lady or a gentleman accepts the outstretched hand she or he involuntarily leans forward a little towards the bars—and in a flash the chimpanzee darts down from his 'bed', grabs whatever he has set his sights on, and carries off his spoils to the swing.

Moritz managed to acquire quite a number of ladies' hats in this way, and would amuse himself by picking off all the trimmings. The team-work of this trio also enabled Moritz to use his unerring aim to knock the hats off the heads of countless ladies and gentlemen if they were not of the kind he could amuse himself with in the cage.

Convincing evidence of the orang-utans' powers of memory was obtained by Hagenbeck about a year after their arrival at Stellingen, when their former owner paid a visit. They both recognized the man with whose family they had been reared by the sound of his voice, before actually seeing him. With animated gestures and raucous greetings they showed their joy at meeting again their former 'head of the family'. They stretched out their dark hands and jabbered away to him, and everyone who witnessed the reunion was convinced that the orangs would have done anything to keep their former owner there with them.

During the Second World War, after a devastating air attack which had severely damaged Berlin Zoo, Dr Bernhard Grzimek looked after a pair of eight-year-old chimpanzees and a five-year-old orang-utan at his home. He had some nice and some nasty experiences with them. He was badly bitten by the male chimpanzee, which, however, later became very docile. From the very day he adopted them he had an opportunity to admire the chimpanzees' artfulness. They had all been accommodated in a double cage in the heated cellar, in which he had previously kept two wolves. Each section had a door leading to the outside, and both were connected by a little sliding door which could be raised by means of a wire from the outside. It did not take the female chimpanzee long to realize that she only had to push up the heavy iron sliding door and crawl underneath it in order to get into the male's compartment!

In his book *Such Agreeable Friends* Dr Grzimek describes many incidents involving these two chimpanzees, named Bambu and Ova, one of which is especially interesting: "Ova removes a pair of tongs from my pocket and tries hard to insert the thin end of it into the lock as though it were a key and then to turn it. Then she forces it in between two of the bars and tries to use it as a lever."

As proof that chimpanzees resemble humans even more in their emotions than in their intelligence, Vance Packard cites the example of a chimpanzee which loved watching a corpulent cook at work; when its keeper closed the door so that it could no longer see, it became extremely angry.

From the 1890's through to the present day, evidence of the intelligence of these creatures has steadily increased, most of it being provided either by experienced animal trainers or by observers at zoos.

Many apes learn by themelves to use tools for various purposes. They also show great adaptability to their new mode of life when in close contact with humans.

Even the possibilities of communication between men and the great apes have been successfully explored, and the sounds produced by a number of different species have been extensively recorded. In 1891–92 the American R. L. Garner tested the learning capacity of capuchin monkeys. At the same time he made an attempt to establish the meaning of the sounds they produced. His conscientious efforts were, however, little appreciated by his contemporaries—indeed, his study of 'monkey languages' at times compromised his reputation as a serious-minded researcher of monkey behaviour. To give Garner his due, it is necessary to consider briefly some of his discoveries.

A capuchin monkey was offered the choice of white and red sweets both having exactly the same taste, in order to ascertain its colour preference. The monkey invariably chose the white sweets. Garner then offered another capuchin white and red paper balls to play with. Again the monkey chose white.

In another experiment he offered his capuchins dishes containing slices of apples and carrots. The amounts in each varied, yet the monkeys always chose the dishes containing the most.

Later Garner carried out a counting test. In a square box he made a hole large enough for a monkey to put its hand through and then withdraw it. The box contained three small metal balls, which the monkey extracted and began to play with. After a while Garner took the three balls away from the monkey and let it see him roll them back into the box. The capuchin's immediate reaction was to take the balls out again one at a time. Having repeated this manoeuvre several times. Garner then slipped one of the balls into his jacket pocket. After finding only two balls in the box, the monkey stuck its hand repeatedly through the hole to locate the third. Finally it began looking for the missing ball in the room. This was the first time evidence had been produced that capuchins are capable of counting up to three. When Garner repeated the experiment using four balls instead of three the monkey did not react to the missing fourth ball.

Garner suspended three small bells in the cage of a capuchin in such a way that they could be made to ring by pulling strings. The monkey watched as Garner gave a demonstration, and then copied him. Two of the bells had no clappers and therefore produced no sound. The monkey soon realized that only one of the bells would ring and concentrated on pulling the appropriate string. Garner then placed a bowl of food in the corner of the cage to tempt the monkey away from the bells. While the monkey was eating, Garner rearranged the position of the bells. Again it did not take the monkey long to discover which string to pull to make the bell ring, and it almost completely ignored the other two.

In Washington Garner made a recording of a whole 'conversation' with his capuchin Pedro, which was normally housed in a communal cage with a number of other capuchins and a spider monkey. The spider monkey, which was twice the size of the others, was the undisputed leader of the group, and kept them under strict control, taking advantage of their good nature and often maltreating young Pedro by pulling his tail and dragging him across the floor of the cage. Garner locked Pedro in a separate cage, where, safe from the clutches of the spider monkey, he now indulged in long conversations with the other capuchins in his chattering and twittering 'language', and it was one of these tirades which Garner recorded. A few days later he played it back to another capuchin, which sat listening attentively throughout. From his observations Garner became convinced that the monkey had entirely different reactions to different parts of the recording.

Garner has also noted down the vocal sounds produced by capuchins which consist quite clearly of vowels and consonants. In doing so he ascertained eight kinds of patterns, containing all the vowels. Shortly afterwards he carried out a test with a capuchin to establish whether it understood a certain sound sequence which he took to be a warning signal. For this purpose he used the services of a thirteen-year-old schoolgirl well known both to him and the monkey, to give the alarm signal. As soon as she had uttered the signal he made as though to strike her and pretended to chase her out of the compound. In the days that followed, whenever the girl approached the monkey it ran away and hid, from which Garner concluded that it connected the warning signal with the girl, whom it remembered as seeming to be a source of danger.

As early as 1820 a French scholar named Pierquin de Gembloux was at work compiling a *Dictionary of Animal Languages*. The chapter dealing with the sounds made by wistitis and their significance was a valuable start to the research into monkey 'speech'. De Gembloux maintained that even the smallest species of monkeys communicated a vast variety of detail to one another, and claimed that he could understand some of their conversation.

The difficulty for de Gembloux was that the monkey sounds only at best approximated to human sounds, and it was therefore extremely problematical to convey them in terms of human phonetics. The same limitations confront ornithologists in their attempts to reproduce in terms of human phonetics the flute-like, chattering, screeching, booming, twittering and other sounds made by birds.

It is true that Garner made use of a gramophone for his 'language research' with capuchin monkeys, but in the 1890's such instruments were somewhat primitive and in no way comparable in quality to those in use today. Garner experienced similar difficulties in 1896 during an expedition to West Africa, where he made use of a phonograph to record the vocal utterances of chimpanzees in the forests. Nevertheless, he publicly declared that he could understand many words of 'monkey language' and could even conduct real conversations with animals. These assertions were not believed by his contemporaries. If recording instruments of present-day quality had been available to Garner it might just have been possible for him to reproduce 'monkey words' accurately enough to convince his opponents.

While a guest at a mission station, Garner built a massive metal cage banded with barbed wire and installed it in a jungle clearing about two miles from his hut. There he listened, often for days and nights at a time, to the voices of gorillas and chimpanzees. He was inspected in his cage by prowling leopards and inquisitive monkeys; but fortunately, the gorillas and chimpanzees, though staying within hearing distance, made no attempt to molest him—which was perhaps just as well for Garner, seeing that the cage, sturdy as it was, would have been unlikely to withstand the brute strength of a gorilla.

Garner returned to Africa for a second trip, lasting several years, during which he claims to have heard chimpanzees 'talking' to one another. After his return to the U.S.A. he tried to teach Susi, his intelligent pet chimpanzee, a number of English words. However clever Susi may have been at inserting toy bricks of different sizes into their corresponding compartments, and however much she may have enjoyed playing games with Garner or with children she knew, she did not react at all to her 'language lessons'. She did not utter a single word, despite all Garner's patient efforts. American newspapers suggested banteringly that he would perhaps have better results if he switched to teaching Susi Latin, a language rich in vowels.

During the decades which followed there were other men who investigated 'monkey languages', including the French zoologist Boutan, who in 1912 brought out a *Dictionary of the Gibbon Language*, and the German philologist Georg Schwidetzki, who in 1931 published a book entitled *Do You Speak Chimpanzee?* The American zoologist Robert Yerkes, director of the ape station established in Florida in 1930 for primate research, has also reported extensively on the sounds produced by chimpanzees, though he himself declares—with refreshing objectivity—that his findings are still incomplete.

Thus while men have gone to a great deal of trouble to plumb the mysteries of 'monkey language', monkeys for their part have shown no interest whatever in learning ours. There are very, very few authenticated instances of monkeys being successfully taught to reproduce human words. One such case involved a pig-tailed macaque whose owner got it to repeat the words 'hunger' and 'Nana'; this particular monkey was later acquired by Leipzig Zoo, where it delighted visitors by begging food and repeating the word 'hunger'. The zoologist W. Marshall visited the macaque several times in the 1890's and heard it pronounce the word quite distinctly. In 1930 there was a 'speaking' orang-utan in the U.S.A. which could say two words. The American zoologist Furness repeated over and over again the word 'Papa' for several minutes each day over a period of six months, until the creature clearly repeated it one day. It took several months before it could pronounce the word 'cup'. These two isolated words were subsequently used by the orang on numerous occasions and in such a way that some people believed it understood the sense of them. When asked, "Where is Papa?", it would run over to its keeper and place one of its hands on his shoulder. If it heard him coming but he was still out of sight it would exclaim with delight, "Papa! Papa!"; and every morning it greeted him in the same way when he entered its cage. When it wanted something to drink it would point to its bowl and say "Cup" with its mouth wide open. It also spoke this word if it found its cup was missing. But, like the pig-tailed macaque, it never extended its vocabulary beyond two words.

The Russian zoologist Nadie Kohts has recorded twenty-three sounds uttered by chimpanzees at Moscow Zoo and attempted to interpret them together with a large number of recurring gestures which are evidently related to various 'words'. One of her conclusions, already observed by many chimpanzee trainers, is that shaking the head is just as much a negative response among chimpanzees as it is among humans. She has succeeded in carrying out short 'conversations' with chimpanzees in their own 'language', always receiving responses from them when she has addressed them in 'Chimpanzee', though she has so far been unable to interpret them.

Much the same thing happens in circuses and zoos between the trainers or keepers and their lions and leopards. They imitate the sounds made by these animals and immediately receive a response. So far, however, no one has been able to interpret these sounds.

In his book *Illustriertes Tierleben,* published as long ago as 1870, Brehm reported in detail on the 'social hierarchy' among baboons and related species. Similar findings have often been made since then.

R. L. Garner observed the existence of a form of social hierarchy among a group of five capuchin monkeys. The strongest of them, Micky, took many liberties with the other four. The second strongest, also a male, while showing respect for Micky, made life rather uncomfortable for the other three. The fifth in the social order had to put up with a great deal from all the other four without ever daring to challenge the *status quo.*

In the monkey house at Leipzig Zoo Professor William Marshall noticed in 1895 that a hamadryas baboon was behaving in a strange manner. This baboon, the strongest member of the whole monkey house, stood first in the social hierarchy. Professor Marshall used to feed them during the summer months with cherries, plums, and other fruits in season, distributing the food in such a way that even the smallest macaques received their share, though he could not prevent the baboon from blustering around to snap up the tastiest morsels. Once all the fruit had been distributed, the baboon would seize the nearest macaque, hold it tightly between its legs, open its mouth with its left hand and extract the fruit stored in its cheek pouches, eating every bit itself.

Alpine rescue dogs at work after an avalanche

112

Lumpi makes himself useful

Photos: Conti Press

Not every Alsatian can perform like this *Photo: Dr Jesse*

back safe and sound to Hamburg-Stellingen, where she was given a period of leave after all her exertions. She later returned to circus life.

During the Second World War, too, elephants played an important part militarily, though not in France as Jenny had done. But they were used on a large scale in the war in Burma.

The elephants used in that war formed part of the six thousand, all fully trained and capable of working on their own initiative, belonging to the Bombay-Burma Corporation, which were scattered over a number of stations in the forests of Burma. Their task was to transport teak to the depots. 'Specialists' among them were put to work in the sawmills, where they displayed considerable team spirit. They were employed as fetchers and carriers, besides being responsible for stacking the sawn timbers. Selected bulls then carried the teak from the mills to the quayside, making use of both their tusks and their trunks. The cows were unable to perform this task, as they have no tusks, but they made themselves useful in other ways. In fact, without them the whole teak exploitation programme would have been impossible, as the bulls were comparatively few.

In 1942 the war spilled over into Burma when the Japs invaded it from Thailand. The situation changed dramatically overnight for the managers of the work stations, the forestry officials, and also for the *oozies* (drivers) and their charges. The Bombay-Burma Corporation, though a private undertaking, immediately placed itself under the command of the State and the Army. The orderly withdrawal of the English civilian population in the face of the advancing Japs was largely due to the unrelenting devotion to duty of the elephants.

The elephants were put to work part of the time on the building of roads and bridges. In this they showed how cautious and distrustful they are: they dislike ground which is not firm and often refuse to walk up a gangway onto a ship, which is why they are usually placed in a special harness and hoisted on board by crane or derrick. Elephants which are not yet fully domesticated are shipped from point to point in solidly constructed crates.

In the Burma campaign it was impossible to induce the

elephants to walk across bridges which, built exclusively for pedestrians, seemed to them too frail to bear their weight, yet they were quite willing, after a few tentative steps, to cross bridges which they themselves had helped to build for the use of army vehicles: it seemed as if they were able to gauge the weight such bridges would bear.

Many elephants and their keepers were later captured and forced to work for the Japs. Thousands of *oozies* managed to escape with their beasts into the jungles of the north, however, where they kept themselves hidden until the British had driven the Japs out.

Only forty-five keepers managed to escape with their elephants across the mountains into Assam. These returned with the 14th Army to Burma along the roads which by then had been made safe for them. With the help of elephants it was also possible to bring food supplies into hunger-stricken mountain areas impassable to jeeps. Because of the exceptionally hazardous condition of the roads in Burma elephants have played an important part since the Second World War in transporting passengers and goods.

To mount their elephants more easily, and also to spare them the difficult manoeuvre of standing up with heavy loads on their backs, the keepers use a method devised over a hundred years ago: the elephants are trained to raise their right or left foreleg and to hold their trunks out sideways, so as to provide two steps up for their keepers. Some are trained to go one step further by raising their keepers gently into the air with their trunks and setting them down comfortably on their necks.

Asiatic elephants have become rare in their old homeland, India. The number of those living in freedom has diminished over vast areas so drastically—because of the population explosion, large-scale rounding up and the earlier massive slaughter of the species for ivory—that for hundreds of miles round such cities as Benares, Bombay, Delhi, and Calcutta there is not a single wild elephant to be seen. Stocks of domesticated elephants have become considerably smaller, and the demand for work-elephants has disappeared. Since maharajas now have to pay an elephant tax, far fewer are kept for ceremonial purposes.

Elephants as a means of transport have been superseded by Bentleys, Mercedes and Rolls Royces.

Nowadays very few people own elephants. Among those who do are authorities in charge of important temples and holy places, and a handful of wealthy lay dignitaries.

From 1947 till the mid 1950's the majority of the elephants which belonged to the previously independent ruling princes of India were purchased by large animal dealers in Europe and the United States of America and by numerous zoos all over the world. Almost all the maharajas of the former native states offered their elephants—"guaranteed tame, well-trained and safe on the roads"—for sale, and the sale catalogues described at length their acquired talents and abilities.

When an elephant was sold to an overseas customer an 'instruction handbook' was supplied with it, so that the new keepers would know what to do with their acquisitions. Their former owners, the maharajas, deserve a special tribute for their continued interest in and affection for their old friends, as even after they had been sold inquiries were made as to their state of health and how they were settling down in their new surroundings. Often details which had been overlooked about their particular qualities were sent on. It is clear that the princes were reluctant to part with their elephants. Even today elephants belonging to Indians are offered for sale "at a cheap price, owing to circumstances". Whether the price really is cheap, however, is questionable, for by the time the foreign purchaser has settled the freight bill and paid for food supplies and an experienced attendant for the voyage, he will have incurred considerable costs over and above the 'cheap' purchase price. But wherever the maharajas' old friends may be now, scattered all over the world in circuses and zoos, their new owners have nothing but praise for them and their intelligence, their readiness to learn, and their docile behaviour.

Nowadays elephants can be obtained simply by placing an order by telephone. If a circus proprietor in South Africa or North America wants to purchase a group of well-trained, intelligent Indian elephants, or the director of Ueno Zoo in Tokyo wants an African bush elephant to extend the range of his elephant house, the interested parties just contact the firm of Hagenbeck in Hamburg-Stellingen or in Alfeld. A significant part of the world trade in large animals is carried out through these channels.

The German trade in large animals reflects a whole century of tradition. As far back as the 1860's Carl Hagenbeck sent out his own animal collectors to Africa to bring back elephants and other wild animals for him.

In 1866 the first large consignment arrived in Hamburg from the Sudan by way of Trieste, Vienna, and Nuremberg. Others soon followed. One of these consignments, consisting of elephants, giraffes, ostriches, and other creatures, was assembled by Hagenbeck to form a 'Nubian caravan', which attracted hundreds of thousands of visitors to the Berlin Zoo, even though at that time it was situated well outside the city in a sparsely populated sector.

Where the Kaiser Wilhelm memorial church now stands there was then a stop for horse-drawn omnibuses, which jogged their way in half an hour to the Potsdamer Platz and in forty-five minutes to the Alexander Platz. Most people went to the Zoo by horse-drawn cab—wealthy people in a landau or their own coach. One day a certain Professor Virchow paid a visit, to see the African collection which had become the talk of the town. The Professor, an authority on wild life, told Carl Hagenbeck that in his view African elephants, unlike Indian elephants, were untrainable. Hagenbeck, however, who had already had a number of years' experience with African elephants, was convinced that they were capable of being trained. He offered to train the five newly arrived African elephants to the point where, within twenty-four hours, they would be willing to take a man for a ride on their backs. Professor Virchow regarded this as quite impossible.

As soon as the celebrated professor had left the Zoo, Hagenbeck arranged for some of the Nubian attendants who had travelled the whole distance with the caravan to mount the elephants and, despite some initial resistance, to ride on their backs. Cajoling them with buns, succulent green vegetables and carrots, the Nubians won over three of the elephants by evening, and next morning the other two followed suit. "It now only remained to persuade them to carry loads instead of men", Hagenbeck recalls. "I ordered some sacks to be filled and bound together in pairs with straps, and these were then hung over the backs of the elephants. The beasts at first disliked the feeling of the loads resting against their flanks, but they soon became accustomed to this sensation also."

When Professor Virchow arrived at the Zoo at the appointed hour Hagenbeck was able to demonstrate to him that African elephants can indeed be trained. The Nubians rode the docile beasts up to the distinguished guest and then rode them away with loads on their backs.

That summer's day in the mid 1870's is in fact one of historical importance: since Hannibal's march across the Rhône and the Alps with elephants the recorded instances of the Romans attempting to domesticate African elephants are very few, and as time passed knowledge of the teachability of this species was lost—so much so that it was taken for granted that it was pointless to attempt to train them. Carl Hagenbeck's hasty training programme showed that it was indeed possible, and this contributed to the establishment some years later of a Belgian training centre in the Congo. This centre, set up at the end of the nineteenth century, was at first situated at Api, near the present-day Congo elephant reserve. It was later transferred to Aru, a border town which separates the Congolese Republic from Uganda, about 75 miles north of Lake Albert.

There are many examples which show how much the tame, docile elephants here responded to kind treatment by their human trainers. One lame bull elephant was taken a long way from the domestication centre, where it had spent most of its life, and set free; but instead of joining up with the elephants that roamed wild in its new surroundings, it made its way straight back to Aru. Another attempt was made to give it its freedom, but that failed too. It always found its way back and hung around the entrance until it was brought in, refusing every chance of liberty in favour of human company, prepared meals and familiar surroundings. It was allowed in the end to stay on permanently at the centre, although it limped, as an 'instructor' of young captured elephants.

The Belgians had a similar experience at Aru with a number

of completely domesticated cow elephants. They had escaped and could not be recovered, despite an intensive search. Yet a few days after the search was abandoned they arrived back at the entrance, trumpeting to be let in, as though their brief flight to freedom had been a bitter disappointment.

The *Station de Domestication des Éléphants* still exists and is now maintained by the central government of the Congolese Republic. The Belgians working at Aru during the disturbances in the Congo between 1961 and 1965 were not molested by the native population, and today fully trained elephants are still transported from there to America and Europe, to find their ultimate home in a circus or a zoo. Some of these elephants are further trained on arrival for performances in the circus ring.

Carl Hagenbeck was an eyewitness to the violent reactions of his highly intelligent elephants to fire, smoke, and loud noises.

On 31st July, 1888, to mark the centenary of his family's business enterprises, Carl Hagenbeck had taken his circus to Munich. Eight elephants had been selected to lead the procession. Four of these were ridden by experienced keepers, the other four by men who knew the creatures well. The noise of the crowds lining the streets made little impact on the elephants, who were used to loud applause in the circus ring and to the clatter of the circus being erected and dismantled. Spectators threw apples and buns to them without their becoming in any way disconcerted, though they only took those which fell immediately in front of them.

In the Ludwigstrasse the Prince Regent took the salute, with the elephants raising their trunks as a mark of honour. On the return journey they had to pass a steam-engine adorned with painted cardboard so that it resembled a fire-breathing dragon. One of Hagenbeck's officials had expressly told the driver and passengers to stop the engine, shut off the steam, and refrain from using any whistles while the elephants were going past. But the driver and passengers, who had been celebrating the occasion with good old Munich beer, ignored the instructions—no doubt because they thought it would add to the fun, as they could hardly have been aware of the disastrous consequences their high-spirited behaviour would have. And so the driver let off steam and blew his whistle just as the eight elephants were alongside his engine. The elephants were already apprehensive of this monster the like of which they had never seen before, and four of them bolted across the street and down through the Odeonsplatz. Panic quickly seized the crowds. Hagenbeck was at first able to control the other four elephants, but then a gang of tipsy rowdies advanced towards them with sticks and knives and struck at them, jabbing them in the trunk. Hagenbeck himself was pushed out of the way by the howling mob and nearly trampled underfoot, and the terrified elephants stampeded.

Mounted troops were called in to head the elephants off, but their gleaming swords only made things worse, and it was not until the keepers finally managed to reach them that the elephants were pacified.

Because of the engine-driver's thoughtless action a large number of people were injured, some of them by being knocked down by the militia's horses. Yet, extraordinary as it may seem, not a single person was harmed by the stampeding elephants.

In every circus the largest elephants are enlisted to help put up or take down the tents when there is a shortage of labour. If in wet weather the tractors are unable to tow the heavily loaded trucks over the soft ground, the elephants are given the job. Their amazing achievements have certainly contributed to the high prestige which German circus owners enjoy on their foreign tours.

During their tour of Austria, Hungary, and Italy in 1930 the Krone Circus often had to cope with prolonged periods of rain, especially on the days of departure. When the circus props had to be taken down overnight, the task was conscientiously carried out by the largest of the circus's twenty-six elephants. Although they had only just finished doing their tricks at the last performance, they willingly allowed themselves to be harnessed and in pouring rain set about hauling the heavy trucks across the soggy ground to the made-up roads. They never refused to work, even in the worst weather. Carl Krone gained the impression that the elephants realized that their help was being depended on.

After Lorenz Hagenbeck had delighted millions of spectators

in various parts of Japan throughout the summer of 1933 with his trained elephants and other performing animals, his whole Noah's Ark was shipped off from Osaka, the last Japanese town of the tour, to China, in a specially equipped German steamer. Pouring rain had completely softened the ground on which the circus was to be erected in the centre of Shanghai. The circus trucks got stuck in the mud up to their axles. It was only thanks to the elephants that the opening performance could begin on time. They did not mind the strenuous work in the rain at all—in fact, they revelled in it, treating themselves to a mud bath, which the wild elephants of Assam, Burma, and Ceylon are so fond of. Hagenbeck comments: "But when our working elephants plunged to their hearts' content into the mire, trumpeting with delight as they wallowed in it, everyone began to feel happy again." Their help was again needed when, because of the streams of visitors to every performance, the tent was in danger of collapsing: the willing beasts stood, unflappable, like a living wall, supporting the front of the main entrance.

"If animals could be awarded medals", Hans Stosch once said to me, after telling me about the way elephants had helped to erect and dismantle circus equipment, "the first to be honoured would be our elephants, which have such keen judgment and are so civilized!"

On countless occasions I have watched elephants in their role as transport workers, pushing heavy circus trucks with their foreheads (protected with leather padding) or in harness hauling their loads, working with precision and never jibbing. The most remarkable thing is that when they enter the circus ring, even just after a really heavy stint, these good-natured animals, with their artistry, are still the stars of the show.

The efficiency with which elephants work, though taken for granted by many people, does in fact presuppose a high degree of mental ability. The precision shown by Hagenbeck's elephants from Hamburg-Stellingen particularly impressed the director of Rome Zoo, Dr Knotterus-Meyer, "when one of them pulls eight to ten loaded lorries behind it as though it were going for a stroll. Or when, at a signal from its keeper, it kneels down to move large lumps of rock with its forehead, then gets up to re-position itself before repeating the action over and over again till the rock is in the right place. Or when it carries girders or tree-trunks, after establishing exactly where to hold them to distribute the weight evenly."

About his own experiences in training elephants he says: "The usual tricks, like playing a barrel-organ, dancing, passing coins to their keepers, which they quickly learn in return for a bun or some fruit, expressing thanks and making requests by going down on one knee and trumpeting—such things elephants have no difficulty in learning."

Further evidence of the intelligence of trained elephants was provided in Hagenbeck's animal park by a female named Ruma. Without any prompting she took to disciplining the less 'civilized' of her companions. The younger ones were roped together with her, so that she could take them on a tour of the park some mornings when there were not many visitors, to get them used to human company. If one of her pupils seemed at all reluctant, Ruma asserted her authority by striking it hard with her trunk. Her corrective treatment never failed to achieve the desired results. The elephants she had instructed became in time as gentle as lambs.

CHAPTER THREE

GOOD CHARACTERS

THERE are any number of stories to illustrate what good memories elephants have, their gratitude for kindness shown to them, and their tendency to bear grudges, sometimes for years, against people who have offended them in some way.

Plutarch tells of an elephant which took revenge on its keeper, who had mixed stones and earth in with its feed, by seasoning his bowl of food with sand.

According to one seventeenth-century account a tame elephant in southern India recognized a soldier, who a few days before had blocked its path and refused to budge, sitting on a river bank with some friends. The elephant got its own back by seizing the soldier round the waist with its trunk, lifting him over the heads of his friends and dumping him in the river. It then lifted him out again before going calmly on its way. The behaviour of this elephant, as described here by Buffon, suggests a measure of revenge tempered, however, with moderation, as if to warn the offender not to take any further liberties.

All domesticated elephants—even the most intelligent—have a weakness for alcohol. For centuries the Indians have exploited this: if they want their elephants to carry out work of a specially arduous kind they encourage them by giving them a taste of millet beer or arrack punch, and the elephants then know that once the work has been done they will be given a full quota. But it is essential for the keeper to keep faith with them, as otherwise they would almost certainly turn on him and show him no mercy.

In the hilly region of Bangalore, in the Deccan, one elephant took its revenge gruesomely on a keeper who had cheated it of its promised reward for services rendered by stamping on his head. The victim's widow, in her grief, offered herself and her children to the enraged beast, exclaiming that she did not want herself or her sons to go on living now that her husband had been killed. The elephant calmed down immediately, as though regretting what it had done. Then it picked up the bigger of the two children and placed him on its neck, to show that it wanted him as its new keeper, and after this it would never allow anyone else to give it orders.

Anyone acquainted with the ways of domesticated elephants will not doubt the truth of this story as told by Buffon. It is quite probable that the keeper who had failed to keep his part of the bargain had already allowed his elder son to ride the elephant and had given him lessons in how to manage it. It is still quite common to see boys in their early teens riding elephants and giving them instructions. As all Asians treat elephants with respect, and children are instinctively kind to them, it is natural that most elephants prefer a young lad as their driver.

Visitors to the Jardin des Plantes in Paris in 1798 witnessed an amusing incident involving one of the two elephants which had been transferred there from the Netherlands. In front of the sturdy bars of their enclosure a sentry stood guard, because both elephants had been captured in the war between the two countries. Knowing how visitors like to feed the animals, the director had put up a notice forbidding the elephants to be fed, and it was the sentry's duty to ensure that this order was observed. Most of the sentries turned a blind eye, but one took his duties very seriously and prevented the public from giving the elephants anything at all to eat.

Parkie, the female, who was more intelligent than her male companion (as is often the case with elephants), quickly realized how unwelcome the presence of this particular sentry was. She waited till he was just passing her, then seized his rifle with her trunk, pulled it through the bars and with one foot bent the barrel almost at right angles. She then tossed the weapon back at him and, for good measure, squirted water at him from close range. For a day or two afterwards Parkie was the talk of Paris.

At the turn of the nineteenth century there was quite a sensation when a tame female elephant, about twenty years old, was unloaded at the East London Docks from a clipper which had brought her from India and was taken to the Duke of Devonshire's residence in Chiswick. A specially heated building had been erected for her in the grounds, and the butler had been briefed on how to handle and feed the distinguished guest.

From the moment she arrived the young lady elephant displayed a warm affection both for the Duke and his butler. She seemed to sense how exceptionally fortunate she was. In India she had been trained to do various kinds of work, and now in Chiswick she seemed keen not to lead an idle life. And so, among other tasks, she carried a huge bucket, usually managed with difficulty by two strapping men, for the butler, who was also in charge of the estate. And, after being shown just once, she pulled the water-wagon along the paths of the estate, and helped to clear surplus trees, simply by leaning heavily against them. At a given signal from the butler she uprooted bushes and even learned how to sweep up leaves and twigs with a besom, which she manipulated skilfully with her trunk.

When the Duke had guests their children were allowed to climb onto the seats strapped to her flanks and be taken for a ride through the park. Previously, in India, she had been given the care of small children and had rocked babies in their cradles. Now, by the Thames, she showed her liking for children again. She let them lead her through the park and looked on them as her protégés. For her services she was generously rewarded by the older children with fruit, buns, and carrots.

The Duke enjoyed telling his guests how she could open unaided a bottle of fruit juice or watered-down whisky, provided the cork was not too tightly fixed. To do this she laid the bottle in a sloping position on soft ground, held it firm with one foot and drew the cork with the tip of her trunk. She then grasped the bottle neatly in her trunk and poured the contents into her mouth. The Duke was always prepared to lay bets with his guests that she would perform this trick without spilling a single drop. As if sensing that both the Duke's honour and her own were at stake, she always succeeded in winning the bets for her owner.

Over a period of years Dr Knottnerus-Meyer made a detailed study of a tame African elephant named Toto. Whenever he appeared with a piece of chocolate Toto would come hurrying along; but if he appeared with nothing the disappointed elephant would spray him with sand. Before allowing his keeper to remove his empty rice bowl, Toto would first insist on making quite sure that there was nothing left in it. If the keeper neglected to show him the bowl before going off with it, Toto would trumpet with annoyance until he did so.

Some of Dr Knottnerus-Meyer's observations about Toto are paralleled by others I made myself some years ago concerning a female Indian elephant belonging to the Krone Circus. Assam was one of the nicest and most docile creatures you could wish to meet. One day she removed the straw hat from the head of a newspaper reporter accompanying me round the elephant house and ate every scrap of it. Dr Knottnerus-Meyer reports: "One thing which gave Toto great pleasure was to appropriate ladies' hats through the bars of his cage, which he did with the speed of a monkey."

Though they daily consume two or three stale loaves and a large quantity of oats, bran, or coarse-ground corn as well as a sizeable pile of freshly cut grass or a less voluminous quantity of turnips or other root-crops, elephants still enjoy an extra ration of hay or straw, so it is not surprising that an occasional straw hat is welcomed as dessert. The director of the Leipzig Zoo has observed that elephants are equally partial to wickerwork baskets and besoms, and have even been known to eat leather briefcases. One elephant made a meal of a rucksack!

Toto demonstrated that he was just as skilful as any Indian elephant at turning on hydrant taps. Walking through the streets of Dar-es-Salaam one day advertising the circus like a well-trained sandwich-man, he manipulated the lever of a water-pump with his trunk and quenched his thirst.

At the Rome Zoo Toto made friends with an Indian female named Greti. One day, having injured his trunk, Toto was confined to his enclosure. To avoid hurting his trunk, his food was placed direct into his mouth and he was given no hay. Greti made up for this by taking across to him from her adjoining enclosure several trunk-loads of hay from her own pile and pushing them through the bars. Dr Knottnerus-Meyer comments: "I was dumbfounded by this Good Samaritan gesture and must confess that I would have found it difficult to believe if I had not seen it with my own eyes."

CHAPTER FOUR

PRACTICAL HABITS

IT is not at all difficult to illustrate the prudence and discrimination shown by elephants even in their ordinary daily routine. For instance, when being transferred from railway trucks to road trucks, they prefer to step straight from one into the other, without a ramp being used. If this is unavoidable, however, they carefully test the ramp to make sure that it is firm enough to take their weight.

Large elephants leave trucks rear first if the doors are lower than shoulder height. They do this by taking up a half-kneeling posture as they squeeze out, snorting and groaning.

When elephants first arrive at zoos it is remarkable how cautiously they inspect the artificial ponds provided for them in their enclosures. They kneel down at the water's edge and gauge the depth of the pond with their trunks. If satisfied they enter it gingerly, probing the steps leading down with their trunks, then they tread circumspectly about until the whole bottom has been thoroughly tested. Only then, after they have satisfied themselves that the pond holds no unpleasant surprises for them, do they abandon themselves to the pleasures of bathing in it.

A couple of elephants, named Dicky and Mary, were successfully reared by the father of Adolf Althoff, a famous circus owner and animal trainer. As small boys, Adolf found his brother watched their progress with passionate interest and soon gained the affection of both of them, as the children indulged them with little acts of kindness and attention. They brought them turnips, cabbage, loaves of bread, apples, and greenstuff—and were amazed to see them trample on them to make them a convenient size for eating. Dicky would place the cabbage or bread in front of his left forefoot and then tread firmly on it until it was broken down into small pieces. This showed that he was 'left-footed', whereas Mary mostly used her right foot.

On Mary's twentieth birthday Adolf's father baked a cake-mix in the shape of a basket, coated it with chocolate, and filled it with lots of ripe yellow apples. Mary enjoyed the fruit, and allowed Dicky to share it with her. She then discovered that the container was also edible, proceeded to crush it gently with her forefoot, and offered a piece to Dicky. After tasting it herself, she trumpeted shrilly and turned her head away. Curiously, Mary did not like chocolate, but Dicky did and he had the whole basket to himself. There were other things which Dicky, in turn, did not like. Once he was offered some heads of celery and a bunch of parsley. He ate the celery, but spurned the parsley. Afterwards he drank three buckets of water one after the other, so thirsty had the celery made him, and he never again accepted celery.

Close observation of circus elephants has made it possible to assess the adaptability and resourcefulness of elephants living in their natural surroundings. Dr Hediger, for example, has shown how readily circus elephants grow accustomed to all the different kinds of noises around them. They settle down to sleep even when men are going past just a short distance away, while tractors rumble by or chains are rattling.

During my travels with circuses, which extended over ten or more years, I have often observed elephants making pillows out of hay and straw and settling down to sleep on them, quite unperturbed by my presence. A cow elephant always stood sentry if the trusted keeper was not present, but if he was they considered it unnecessary to post a sentry, unlike wild elephants, which always do so at night.

When a monkey wants to scratch itself it can reach any part of its anatomy with either its hands or its feet. An elephant, on

Photo: Professor Dr B. Grzimek

Karl Krall with his 'reading' and 'counting' Arab stallion Zarif

Wilhelm von Osten with Clever Hans

Photo: Historisches Bildarchiv Handke

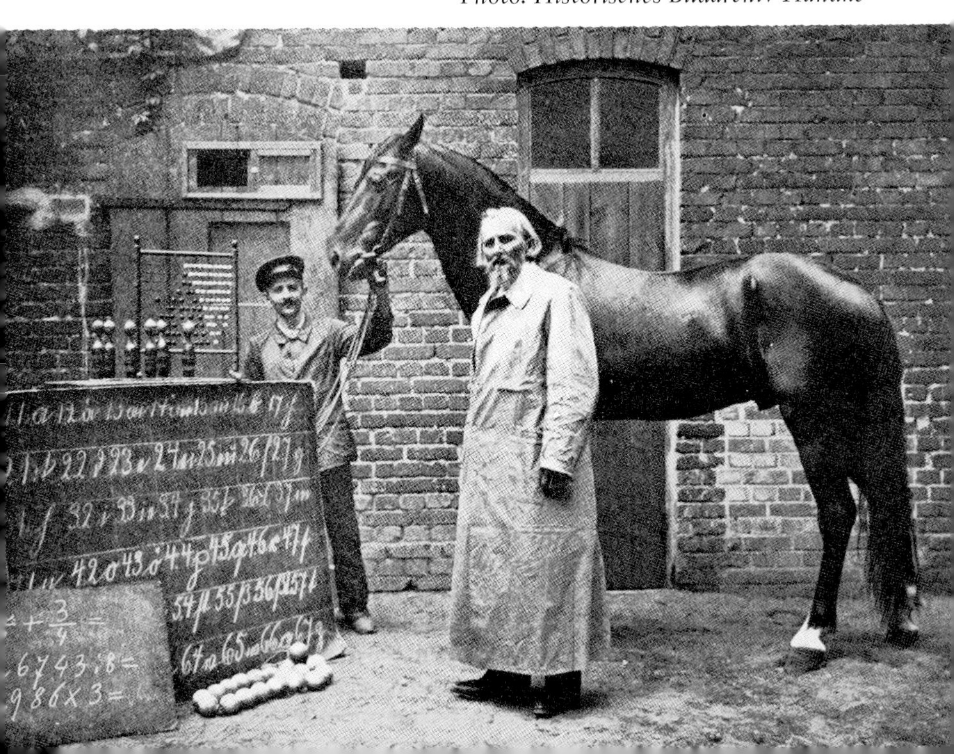

Photo: laenderpress

Horses in training must have quick understanding and a good memory

toothpick. Once when she was let off the chain to see what she would do she made straight for a bottle of Malaga and, pulling out the loose cork, she washed down the roast dinner she had just eaten. After a few hearty gulps she wiped her thick lips with the back of her hand and put the bottle back where she had found it. Vosmaern does not mention whether she left the cork lying around or replaced it. The trainer Faszini, however, observed that his chimpanzees often re-corked bottles of fruit juice, of which they were particularly fond, that had already been opened, and even completely empty bottles. He then experimented with bottles of lemonade whose tops had special lever devices. These the chimpanzees removed and replaced with no trouble at all. Whereas Vosmaern's orang-utan drank straight from the bottle, Faszini's chimpanzees always poured their drinks into mugs, glasses, or cups.

Brehm, in his *Illustriertes Tierleben*, quotes the story of a German sea-captain named Werner, who attempted to take back alive to Germany a male orang-utan—an event which at the time caused quite a stir. Already completely tame, Bobby, as he was known, was allowed the run of the ship and enjoyed clambering on the yards in the tropical waters of southern Asia so that he could look down on the crew working below. At lunch-time he sat at the captain's table and made sure he was not served short rations, especially when fish was on the menu. But he ate with his fingers and could not get used to a spoon. Cups or bowls of soup he would put to his mouth, thrusting his lower lip forward to drink it. After lunch he was always given a glass of wine by the steward. Invariably he would sniff it before pouring it into the ladle-shaped space made by protruding his lower lip, from which convenient container he would slowly gulp it down.

Although accustomed to the superior food served to the officers, Bobby did not find it beneath his dignity to join the crew for an occasional meal, provided it was something he specially liked. Every Tuesday and Friday, when the crew had sweetened sago flavoured with ground cinnamon, Bobby would honour them with his presence and be given a generous helping.

Bobby also had a great weakness for flour. He knew just where to find the flour barrel in the ship's galley, and as soon as the cook turned his back for a moment would reduce its contents by several handfuls. The crew always knew just where their normally auburn-haired friend had been, because he had the habit of wiping his hands on his head. Then they would call out to one another: "Bobby's been powdering himself again!" Apart from the food already mentioned, Bobby received a daily ration of two coconuts for health reasons. Having pierced the hard shell in exactly the right places with his still harder teeth, he first drank the sweet milk with relish, then dashed the nut to pieces on the nearest iron object and munched happily at the tasty flesh.

Bobby's favourite playmate was one of the macaque monkeys, smaller than himself, which were also on board. For a time these monkeys were not at all popular with the crew, because of the extra work caused by the tricks they used to get up to. Like the other monkeys, Bobby's favourite, when threatened by an exasperated member of the crew, would rush over to him screaming and cling to his hairy chest. Bobby would then scamper off into the rigging with his protégé and not venture down on deck again till he was sure that there was no longer any danger.

The boat had already rounded the southern tip of Africa when Bobby, who stayed most of the time below deck to keep cool, fell victim to an incident which began by being comic but ended tragically. Having somehow managed to find his way into the wine-store, he had opened a bottle of rum and drained it to the last drop. For a quarter of an hour, while the crew anxiously kept their distance, he behaved like a man completely intoxicated, dancing about with no control over his limbs, then he keeled over. Several hours later he came to his senses, staggered over to his bunk and lay down. He had a temperature and remained in his bunk for a whole fortnight until, in spite of all the care lavished on him by the captain, whose medicines he took without complaining, Bobby died.

In 1927 and 1928 the importation of orang-utans suddenly took an upward swing. This 'explosion' was closely associated with Messrs Ruhe, one of Germany's leading firms dealing in wild animals, and a Dutch wild-animal collector, van Goehns, who lived in Sumatra and was under contract to that firm. In 1926 van Goehns had sent Messrs Ruhe a male orang-utan, which was subsequently acquired by Professor Brandes, director of the Dresden Zoo. The Professor had an entirely new type of enclosure built for this prize specimen, which he named Goliath, with lots of bars and poles set at various heights. Goliath survived there for over two years. His weight when he died was 132 lb. and his height 136 cm. (4 ft. 5½ ins.). When he stretched his arms sideways, as he often did on the bars, the distance between his finger-tips was 256 cm. (8 ft. 5 ins.), compared with 170–180 cm. (5 ft. 7 ins.–5 ft. 11 ins.) for an average adult human.

Goliath, on arrival in Germany in the autumn of 1926, had stayed for a while in Alfeld near Hanover, where Messrs Ruhe had their offices, before being transferred to Dresden by Professor Brandes. He was a welcome—and profitable—addition to the very small number of orang-utans then in Europe; but even more exciting than his safe arrival was the letter which arrived with him from Sumatra. In this van Goehns declared, with typical Dutch matter-of-factness, that if Messrs Ruhe required any more orang-utans he could soon supply them—Herr Ruhe need only specify the number required. Bearing in mind how scarce the species was in Europe at that time, it is not difficult to picture Herr Ruhe's delight—tempered, however, with scepticism. If this breathtaking offer was as good as it sounded, it would indeed be a dream come true. Perhaps only the specialists can still remember how sensational van Goehns's offer was at the time.

The offer turned out to be neither a poor joke nor a casual overstatement. In April 1927 a fast steamer with twenty-five fully grown orang-utans on board was racing through the Red Sea towards the Suez Canal. During the voyage a happy event took place: Suma, one of the orang-utan passengers, gave birth to a son.

At Rotterdam the precious cargo was transferred from the steamer into two goods trucks coupled to an express train, and the following day all the orang-utans arrived safe and well in Alfeld.

So many directors of zoos are said to have converged on the

small town that there were not enough hotel rooms for them all. Nor were there enough orang-utans. Twenty-four of them went to the zoos in Amsterdam, Berlin, Düsseldorf, Frankfurt, Königsberg, Leipzig, Munich, and Nuremberg, while the infant born on the voyage was taken with its mother by Professor Brandes to Dresden. And so Goliath, already resident at Dresden Zoo and the first fully grown orang-utan ever to be there, suddenly found himself with a family. An account of the happy relationship which developed between Goliath, Suma, and her baby (who was given the name Buschi), and later between Suma's next mate and their resulting offspring, was given by the Professor some twelve years after the arrival of Suma and Buschi at the zoo. The numerous descriptions of orang-utans in his book prove that no two are alike. Just as there are quite definite facial distinctions between one chimpanzee and another—or between one human being and another—so there are between orang-utans.

The twenty-three apes from Sumatra which arrived at Alfeld in August of the same year were exported to the U.S.A. as a job lot in accordance with a contract signed with a North American firm. In the following year, when the next consignment arrived, consisting of twenty-four orang-utans, the requirements of other European zoos could at last be met. But first the orang-utans had to be sent to the acclimatization zoo situated near the small town of Cagnes-sur-Mer, not far from Nice.

Many apes and monkeys were later to pass through this acclimatization zoo, including hordes of baboons, but from 1928 onwards only very few orang-utans were housed there. The Dutch governor in Batavia (now Djakarta) on the island of Java in 1929 prohibited their export. Since the ones still living in the north-west of Sumatra, in the thickly wooded region of Atjeh, were placed under national protection—still strictly enforced by an Indonesian police detachment stationed at Medan—orang-utans are nowadays seldom exported and even then only by special permission of the government. Exports from Borneo, where they are found chiefly in a marshy region on the west coast and in the wooded uplands of the areas inhabited by the Dyaks, are likewise dependent on licences issued by the government. Every precaution is being taken to save the surviving orang-utans of Borneo and Sumatra from becoming extinct and thus suffering the same fate as the Père David's deer.

It is an easy matter to study the habits of chimpanzees and orang-utans in a zoo and the versatility of smaller species of monkeys in one's own home. But it is rather more difficult to make a detailed study of them in their free-range habitats and natural surroundings. Many surprises are in store for even experienced observers when they look at their subjects in the wild.

English farmers on the Waterberg in Northern Transvaal set up in the late nineteenth century a kind of natural reserve for baboons. Certain fields were set aside for their exclusive use, and they could feed on the maize and other cereals planted there for their benefit. For years the chacmas (grey baboons) were allowed to remain in these preserves, as a result of which a mutual trust grew up between them and the white settlers and their negro labourers. These men were looked upon by the baboons as friends, and were even allowed by them to walk among their hordes without fear of ever being attacked.

Planters who were out and about at night discovered that crowds of baboons would accompany them to protect themselves against leopards. Baboons are frightened of leopards only at night; during the day they will attack them *en masse* and have been known to kill them.

In his book *My Friends the Baboons* Eugene N. Marais describes an encounter with a troop of baboons living in the wild which were engrossed in playing games with a number of African negro children. Half a dozen boys and girls were enjoying themselves some distance from the village, making primitive animal figures from fresh clay, as children in Europe do with plasticine. Europeans are not likely to have around them any animals other than cats and dogs, but these negro children were being actively helped by half a dozen young baboons. Some distance away the parent baboons kept watch and looked on approvingly. The whole scene as noted by Marais is reminiscent of similar descriptions by contemporaries of the African explorer Heinrich Barth a hundred years ago. It was observed then that negro children lived in close community with young monkeys, which went in and out of the huts as they pleased. Accounts from travellers to India in the nineteenth century contain similar reports; there country children were often seen contentedly playing with the young of hanuman langurs, which wandered around the villages at will.

According to some planters, a number of African negro children were seized and carried off by baboons, though they came to no harm, as the baboons regarded them as living playthings, in much the same way as children regard dolls. Sometimes, however, there were anxious moments, such as when the baboons climbed into the treetops or onto the roofs with young babies in their arms.

Marais describes also, by way of contrast, how baboons looked after infants with the full consent of the mothers. The presence of these simian baby-sitters quickly put a stop to the noise of even the most persistent cry-babies. Often the women called on their baboon friends to help get their children off to sleep. Afterwards the baby-sitters would find a saucer of milk awaiting them as their reward.

Baboons also lavish their affection on the young of animals in no way related to them as species. There is, for instance, the case of a billy-goat which was reared by a female baboon and lived as a sort of mascot in the midst of the troop. It is also known that they enjoy having rabbits, and sometimes even cats, as playmates.

The behaviour of manlike apes is similar to that of baboons in this respect. Brehm noted several instances of chimpanzees playing with long-tailed monkeys and mangabeys, and even with dogs and goats. Cherry Kearton's chimpanzee Toto made friends in Nairobi on the head forester's estate not only with his children but also with a tame hyena. The hyena had no objections to the young chimpanzee going for rides on its back.

The female chimpanzee Petra, reared by Miss Lilo Hess, had a domestic cat as her first playmate, then later a sheep and a young ant-eater. The chimp and the ant-eater had their meals at the same table, and both drank from cups, but in different ways: the ant-eater licked the liquid from the cup with its long tongue, while Petra held her cup in her hands and drank from it.

Various attempts by wily baboons to free others captured by hunters are described by Carl Hagenbeck in his book *Beasts and*

Men. A number of baboons, mostly males, tempted by some millet which served as the bait, found themselves caught in a trap. After their captors had taken the usual precautionary measures the baboons were placed in transport cages, whereupon

> whole hordes of baboons made their way to the seriba (collecting station), clambered up the palm-trees and called out in unintelligible sounds to the captured animals, which replied with wailing noises. The conversation ultimately degenerated into an ear-splitting cacophony. One day a particularly spirited member of the troop sprang over the thorn hedge into the camp and ran over to the cage in which perhaps his father, brother, or uncle was.

But his efforts were unavailing, and he was driven off by the African hunters.

In his book *Monkeys in the House* (1951), Bernhard Grzimek tells about a female rhesus monkey which was "decidedly nice and affectionate" towards him. Compared with the three rhesus monkeys which he later kept, he comments: "Never will any of them become as tame as my Rhesi used to be, unless I take it away from the others and keep only it."

Grzimek touches here on an important problem relating to the quick domestication of animals. Young monkeys should be reared on their own if it is intended to keep them as pets. The same applies to cage birds and other house pets.

In the case of rhesus monkeys and of other macaque monkeys, the members of a family group or of a troop band together against man. Every zoo director has had experience of the difficulties involved in catching any particular member of a monkey community in a cage, as practically every one of them is a past master at loosening the running noose thrown over its head and pulling it back over its head to free itself.

If, however, one finally succeeds in attaching a collar and chain to a macaque not yet quite tame, one may well discover to one's surprise that the creature is intelligent enough to accept the inevitable: "It becomes quite withdrawn and docile, and takes hold of the chain with its left hand so that the loop does not hurt its neck, and walks beside me as meekly as a dog" (Grzimek).

Similar behaviour was noted with a patas (red monkey) which was made a house pet. This monkey, at the time only half grown, later grew so used to its collar and chain that it would hurry over to the door of its cage whenever a member of the family arrived with them. It knew from experience that the next hour would be full of fun and variety and that it would be pampered by being allowed to sit on someone's lap and be combed and brushed.

It is much the same with capuchins and woolly monkeys. If they are shown their collar and chain they will come over to the edge of the cage for them to be put on.

For quite a long while I kept a male squirrel monkey in a cage which gave it plenty of room for movement, though it was also allowed a good deal of freedom around the house. Atma was very tame, liked sitting on my shoulder, and enjoyed grooming me as monkeys do. He obeyed—up to a point—when I called him, though on one occasion he did knock over my india-rubber plant, because in jumping from the curtain-rod onto it he had overestimated its stability. He would never allow anyone to attach a lead, yet he did not mind being taken all round the house, and there was never any fear that he would run off. He would always remain sitting quietly on my shoulder. When I gave Atma away later he still adamantly refused to wear a lead, but he was just as faithful to his new owner as he had been to me, and even more so after he was given a female of the species to keep him company.

All this goes to show that the adaptability of the smaller monkeys varies a good deal from species to species and from one individual to another. Not every owner of a monkey can take his pet around on a lead as one can with dogs. Yet a monkey reared on its own, provided it gets used to human company during its early life, almost invariably becomes quickly tame. An animal photographer who has kept lots of monkeys as pets usually needed less than a week to make a young spider monkey or woolly monkey feel at home with him. In the case of other monkeys reared on their own but not brought into contact with humans until they were about two years old the period required to make them tame was rarely more than a fortnight. With capuchins and squirrel monkeys aged less than a year the ornithologist Franz Mönig made the observation that after six to eight days they came to regard him as their 'big brother' with superior intelligence.

Both anthropoid apes and monkeys further down the evolutionary scale fear nothing as much as snakes. The fact is often mentioned in reports about monkeys. The chimpanzee belonging to the Kearton brothers was panic-stricken at the sight of the sloughed-off skin of a puff-adder. The female chimpanzee Petra was terrified when she saw the snake-like coils of cable attached to a photographer's projector. I myself have shown my squirrel monkey Atma illustrations of snakes in a book and every time he recoiled in terror. The same thing happened with a spider monkey belonging to an animal lover in Düsseldorf. It snatched at coloured pictures of beetles and small birds, which many species of monkeys regard as delicacies, was indifferent to illustrations of fish, and even of cats and dogs, but shrank in terror at pictures of adders and other poisonous snakes, jumping onto his master's shoulder for protection in case one of them should suddenly spring to life from the pages and dart out its tongue at him.

Baboons kept in the house as pets are just as capable of opening locked doors as orang-utans and chimpanzees are. Elisabeth, Duchess of Montgelas, owned some African baboons and an Asiatic bonnet monkey which greatly enjoyed having a bath. They all understood how to manipulate the lever that controlled the shower apparatus. When bathing in the river they liked to hold on to the tails of her pet dogs and be pulled along through the water. Whenever they got the chance they would go through the pockets of clothes left lying around on the bank before taking all the items of clothing they could find up into a tree!

In her book *Tiergeschichten* (Animal Stories) the Duchess describes a drinking session which her monkeys once had. They became quite tipsy and began searching eagerly for fleas on their mistress; they then set about terrorizing the dog which lived in with them, draining the glasses as they cavorted around saluting one another in true military fashion. But they did no real damage; only a few flower vases and trinkets were broken. The next day they all slept rather later than usual.

For a while the Duchess kept a female baboon on a chain in the stable. Baboccia, as she was called, would often unfasten the spring-hooks when she thought nobody was looking.

In her book *Exotic Wild Animals in Captivity* (1928) the Duchess claims that Baboccia's level of intelligence was seemingly on a par with that of anthropoid apes. There is undoubtedly some degree of exaggeration in the following assessment: "All the intelligence tests at present being taken by the academic chimpanzees at Berlin Zoo would have been mere child's play to my Baboccia. . . . She could open cupboards and padlocks if given the chance to find the key that fitted."

Baboccia knew exactly in which drawer there were sweets, and would not touch drawers from which she had taken the sweets only a short while before. She could distinguish pieces of bread wrapped in paper from similarly wrapped stones, simply by sniffing at them. To make doubly sure, she would throw the wrapped stones onto the ground. "She could also differentiate between nuts filled with sand and stuck together again and real ones which had not been tampered with."

Many zoos in Europe and elsewhere have monkey islands of various sizes, surrounded by water and high, smooth, vertical walls. One of the largest monkey islands is at Brookfield Zoo in Chicago. In October the monkeys there are rounded up to be transferred to their heated winter quarters. A few years ago the baboons defeated every effort on the part of the keepers to capture them by taking refuge in the trees. In desperation a solidly constructed trap was installed, which looked rather like an outsize kennel. At the front there was a trap-door which could be sprung by a length of rope concealed under leaves and hence invisible to the baboons. Inside, against the back wall, a mirror was fastened on which a bunch of luscious-looking bananas had been painted. Just in front of the mirror a real bunch of bananas was placed. In this way several baboons at a time were captured. The painting on the mirror showed traces of tooth-marks, indicating that the baboons had been deceived into thinking that the painted bananas were real. It was quite clear that these monkeys, living as they did in close proximity and on terms of mutual trust with human beings, did not have the same caution as the baboons roaming freely over the uplands of Natal, in South

Africa. There the planters living in the neighbourhood of Durban got to know their cunning in no uncertain fashion.

These wooded slopes are inhabited by large numbers of baboons, which trespass onto the plantations to reap what the planters have sown. On one of the plantations some of them even broke into the house and made a thorough inspection of the kitchen and larder. The infuriated owner had to use physical force on them before he could get them to leave. The next day he built a large monkey trap, and placed several bunches of bananas in it. That same night he caught and shot no fewer than twenty baboons. The following night only one victim was taken, while in the fields round about a whole swarm was feasting on the crops. Evidently news of the risks associated with the bananas had spread, for the fruit remained untouched the whole of the third night, even though the baboons were so close to it.

As in ancient times, sacred (hamadryas) baboons, and occasionally dog-faced baboons too, are still taken as young animals from their slaughtered mothers in Ethiopia and Somalia and reared by the Danakil and the Gallas, so that when fully tame they can be sold to the itinerant monkey trainers or to wealthy members of the Emperor's court. In the 1940's one of these Amharic princes trained a number of baboons to hold candlesticks containing lighted candles. They were to act as torchbearers at a feast, and were to stand on stone pedestals. The plan was that after a certain time, provided they had behaved themselves well, they would receive some food. Behind each baboon stood its keeper. From their vantage-points the monkeys could see the assembled guests tucking into their food, and they were overcome by a temptation too strong to resist. Throwing the candlesticks to the ground, they rushed over among the guests to take their pick of the food; some helped themselves to fruit, others to roast mutton. No harm came to any of the guests, however, for after they had had their fill the baboons allowed themselves to be taken away quietly by their handlers, back to their cages.

CHAPTER EIGHT

CHOICE AND CARE OF PET MONKEYS

MONKEYS, especially small ones, are still kept as pets, though not to the extent they were in ancient Rome. The tricks these artful creatures get up to never fail to amuse and entertain. Naturally their owners realize—or should realize—that unlike cats and dogs they cannot be given the run of the house, not even the smallest ones.

The large manlike apes do not qualify at all as house pets: because of their great physical strength they are too dangerous, besides being more sensitive than any other primates to variations of temperature, and in any case too costly to buy and maintain. Even so, there are animal lovers who will pay a lot of money for a young chimpanzee.

Baron von Fircks, some forty years ago, reared an ape at his home in Copenhagen. A plantation owner in Cuba in the 1930's kept a monkey ranch which included orang-utans and chimpanzees as a hobby. An American couple reared a young chimpanzee with their own children, treating it as a member of the family with equal rights. Travellers on zoo quests and writers of animal books have also reared young apes. But such cases are nowadays rather exceptional.

The purchase of even a medium-sized monkey needs careful thought. Baboons, for instance, should only be kept in the home while young. Although it is true that females remain more docile than males as they grow older, they develop such great physical strength that they can be a menace. Older macaques, long-tailed monkeys, and mangabeys can be recommended as house pets only with certain reservations. They are better off in a zoo, where they can live in family communities and have more freedom of movement, which is essential for their well-being. For these reasons the majority of people who keep monkeys as pets choose the smaller and less dangerous kinds.

Perhaps the most important consideration, when deciding which species to recommend as house pets, is how much space is available. If a cage 4 ft. high × 18 ins. × 18 ins.—a size easily obtained—is installed, the best choice would be from among the extensive range of marmosets and tamarins listed on page 96. These are all very small species, and there are many others which are closely related; they differ from one another not so much in size as in coloration and minor variations on the head or neck. Only a limited degree of intelligence can be expected from any of them, but what they lack in intelligence they more than make up for in affection. Their utterances consist of a kind of twittering interspersed with singing and chirping sounds somewhat reminiscent of the tone quality of woodland birds. After having one of these small monkeys for a while it is possible to distinguish various vowels and consonants, though they are all at such a high pitch that they sometimes go beyond the range of human audibility, especially with older people.

All marmosets and tamarins need a mixed diet of meat and greenstuffs. The insects and tiny birds which form part of their natural diet in the wild can in captivity be replaced by scraped meat, chopped raw liver, yolk of egg boiled or raw, and meal-worms.

If reared individually, these squirrel-sized monkeys are quick to recognize and show affection for their owners, usually greeting them with gentle sounds.

Like dogs, they get to know the footsteps of various members of the family. If allowed to run about freely in a room for hours at a time, there is nothing they like better to show their affection and gratitude than to climb up on to their master's shoulder and be petted by him. Even if two are reared together the degree of affection for their provider is still high, though they tend more to take refuge in their cages if a stranger enters the room.

All monkeys which are suitable as house pets are, in their wild state, compulsive nest-robbers and hunters of small song-birds. However well fed they may be in captivity, this trait remains with them, and it is therefore very important not to keep a pet bird within their reach.

As companions about the house, the best choice is the capuchin monkeys, the most intelligent of monkeys except for baboons and of course the manlike apes. Capuchins are extremely intelligent and have an extensive range of sounds.

The monkeys specified in the table on page 97 require a cage about 7 ft. 6 ins. in length and at least 5 ft. high. Squirrel monkeys up to two years of age and capuchins up to a year and a half can make do with a smaller cage with an area of 20 ins. × 20 ins. Older squirrel monkeys and capuchins would, however, be too cramped in such a cage. The monkeys shown in this table, all of similar size, become affectionate and tame very quickly if reared separately. Anyone who acquires one should, as soon as the 'settling-in' period is over, accustom it to a harness and chain. The harness is similar to that worn by dogs, to save its neck from constriction. Monkeys should not, therefore, be fitted with an ordinary dog-collar, as they are too frisky and could easily be strangled. Any monkey the size of a cat will quickly take to a lead if reared on its own, because it is intelligent enough to associate wearing it with going for a walk or some other pleasurable diversion.

Outings with a monkey should only be made if one has a garden of one's own, as taking it through the streets could lead to incidents. Even a tame monkey will bite if it is frightened by a man, a dog, or some unforeseen happening. Because of this children should be discouraged from taking a monkey out for a walk as they would a dog.

It stands to reason that in a room where a monkey is allowed freedom of movement anything breakable, such as flower-pots and vases, should be removed so that the monkey can indulge its skill in climbing and jumping without causing any damage. On account of its insatiable curiosity all cupboards and drawers should be kept locked. With the exception of the miniature marmosets and tamarins, every species of monkey will delve and pry wherever it can, turning everything topsy-turvy in the process. Of the three species described in the table on page 98, each of which differs widely from the others in appearance, character, and disposition, only the woolly monkey, within certain limits, is free from these tiresome characteristics and because of this is especially popular.

The monkeys listed in the tables do not by any means represent all the species suitable as house pets. Even among the African mangabeys there are eligible species such as the crested mangabey, and a number of the smaller species of Asiatic macaques have been reared successfully.

Monkeys cannot match dogs in reliability or devotion, however affectionate they may appear. A trained monkey can, it is true, carry a basket of provisions just as well as most dogs can, but unlike dogs they tend to steal anything edible they can lay hands on.

One should be careful not to overestimate the capabilities even of manlike apes in comparison with those of other animals. In this connection Dr Ludwig Heck, in his book *Bobby the Chimpanzee and Other Friends of Mine* (1931), comments:

> It must of course be taken into account that monkeys, in their physical execution of whatever they are mentally equipped to do, are greatly helped by their man-like hands, and the question ought perhaps to be asked whether for example dogs would not be able to accomplish more if, instead of their clumsy paws, which serve only to enable them to walk and scratch, they possessed the incomparably more apt prehensile hands of monkeys.

Popular Small Monkeys (Marmosets and Tamarins)

Name	*Distinguishing features*	*Where found*	*Remarks*
Lion Marmoset (= Golden Marmoset)	Reddish-brown fur, 'lion's mane'.	The rain forests of south-east Brazil south of the Amazon.	They become very affectionate and trusting.
Pinche	White-tufted head; back brownish-grey, chest and belly white.	The rain forests of Central America, Ecuador and Colombia.	Good jumpers; they need a predominantly meat diet.
Wistiti	Tufts of white hair on the ears. Colour of fur varies greatly from grey to reddish-brown.	The rain forests of the south-east coast of Brazil south of the Amazon.	Good climbers and runners, seldom jump; they remain tame and affectionate.
Silky Tamarin	Long, thick, silky fur, rust-brown to white.	Tupi rain forests south of the Amazon.	Skilful jumpers. Quarrelsome with other members of the same species. Sometimes mistakenly identified with the Wistiti.
Silvery Marmoset	Silky silvery-white fur. Black tail. Reddish-yellow face.	The rain forests south of the Amazon from the Atlantic to the Andes.	Excellent jumpers.
Pygmy Marmoset	Mane, round head. Greyish-green fur.	The rain forests in the upper reaches of the Amazon, Rio Napo and tributaries to the foot of the Andes.	Skilful runners and climbers, seldom jump.

Popular Cat-size Monkeys

Name	*Distinguishing features*	*Where found*	*Remarks*
		Central and South America (*'New World' Monkeys*)	
Capuchin Monkey	(a) Weeper, olive-green to dark brown. (b) Black-and-white Capuchin, with white shoulders. Both these species have a prehensile tail.	The jungle north and south of the Amazon. The jungles of Central America and Colombia.	When happy they utter whining noises, when excited they chatter. All species are quickly tamed when young, affectionate, easily taught. Weepers become difficult as they grow old. Mainly plant-eaters.
Squirrel Monkey	Upper parts olive-green, lower parts lighter. Bright pink face with a black mouth. Tail not adapted for grasping.	From Guiana to the foot of the Andes and southern Central America. Other species south of the Amazon as far as Bolivia and Paraguay.	Good climbers and jumpers. Remain as a rule tame and affectionate, even when old. Not nearly so easily taught as the Capuchin Monkey.
		Africa (*'Old World' Monkeys*)	
Green Monkey	Upper parts dark green, lower parts yellowish-green, face grey.	Plains and wooded country throughout Africa south of the Sahara.	Females remain tame even when old, males become difficult.
Red-nosed Monkey	Olive-green fur; blue face with bright pink circles round the eyes; parts of the mouth and nose also bright pink; cheek and chin tufts white and green.	Nigeria to northern Cameroons and on the island of Fernando Po.	Males and females remain as a rule tame and affectionate, even when old.
White-nosed Monkey	Upper parts dark-olive, lower parts light grey to white. Face black with white tuft of hair on the nose.	Jungles from Senegal to Uganda.	Like the other two African species, is easily taught, at times just as obstinate, but less obedient.

Larger Monkeys kept as Pets

Name	*Distinguishing features*	*Where found*	*Remarks*
Patas Monkey	Reddish-brown fur, white moustache and chin tuft. One of the long-tailed monkeys, but equally at ease on the ground.	Plains and wooded country of south of the Sahara from Senegal and the Congo to Somalia and Tanzania.	When young docile, affectionate, and teachable. From the age of five more difficult, though the females remain mostly tame and obedient.
Spider Monkey	Varies from brown to black. Its long prehensile tail serves as a 'fifth hand'.	The jungles of Central and South America. Subspecies: Coaita, from the jungles of Brazil to Bolivia and Paraguay.	If tamed when very young they usually remain docile and affectionate when old, but are sometimes liable to be irritable and bad-tempered.
Woolly Monkey	Fur varies from brown to dark brown or almost black. Like the Spider Monkey, has a long prehensile tail which serves as a 'fifth hand'.	Commonest in the jungles north of the Amazon, less widely distributed south of the Amazon.	Friendly and usually quiet house pets, sometimes given the run of the house with little supervision. Sociable towards dogs and cats, provided these have been reared with them.

PART TWO
ELEPHANTS—WILLING AND ABLE

CHAPTER ONE

OLD FRIENDS

SURPRISINGLY, most people are not frightened of elephants, in spite of their massive size and strength: on the contrary, people like them. The reason may be that, once tamed, elephants are among the most obedient and loyal of all creatures. In fact, they form an attachment to their masters which may last a lifetime. This happy relationship between men and elephants goes back thousands of years. There is evidence that our ancestors tamed the mammoth and the mastodon, two species of elephant now extinct, and that mastodons—but not mammoths—were put to the service of man. Anyone who is well acquainted with the 'civilized' elephant of southern Asia or has seen the African elephant at the circus will need little convincing that mammoths and mastodons were not just spoils of the chase but man's docile companions as well, though nowadays elephants in Africa are regarded by many native tribes as animals to be hunted. The southern Asiatic elephant too, in spite of prohibitions, is still looked upon in the jungles of Further India and Borneo as a source of meat supplies. Yet their primitive forebears, because of their intelligence and loyalty, were for thousands of years much more to man than mere helpers. In intelligence elephants rank almost as high as the big cats.

How wild elephants are captured with the help of elephants that have been tamed was described in detail by the Greek writer

Arrian, who was born in Asia Minor and served for a while as an officer in the Roman Army and also as a government official in Gaul. His account varies slightly from those of Buffon's contemporaries, but is nevertheless very informative. In his history of Alexander the Great's Asiatic expedition (*c.* A.D. 160) and its supplementary account of India known as *Indica*, Arrian reveals a precise knowledge of the aptitude for learning of the domesticated Indian elephant. The use made of these animals for breaking in the captured wild elephants confined in large enclosures is most impressively described:

> The inhabitants on hearing the news that the wild elephants are caught in the enclosure mount the most spirited and at the same time most disciplined elephants, and then drive them towards the enclosure, and when they have driven them thither they do not at once join battle, but allow the wild elephants to grow distressed by hunger and to be tamed by thirst. But when they think they are sufficiently distressed, then they enter the enclosure; and at first there is a fierce battle between the tamed elephants and the captives, and then, as one would expect, the wild elephants are tamed, distressed as they are by a sinking of their spirits and by hunger. Then the riders dismounting from the tamed elephants tie together the feet of the now languid wild ones; then they order the tamed elephants to punish the rest by repeated blows, till in their distress they fall to earth.

The methods of taming these creatures, by use of force on the part of 'collaborators' (the elephants already tamed) against their own kind from the jungle, have not changed so very much in two thousand years.

Buffon, the eighteenth-century French zoologist, has much to say about the teachability of elephants in his *Histoire naturelle, générale et particulière*—for example:

> Discounting man, the elephant is by far the most imposing creature in the world. It approaches man in its perceptive faculty—in so far, that is, as matter can approach spirit. At the same time it has a capacity for being taught similar to that of a dog, and, like it, can show gratitude and very strong affection; it readily accustoms itself to man, is kept under control through kind treatment rather than by force, and serves man

readily, faithfully, and intelligently. Consider how, unaided, it can move machines and carry loads which six horses would be unable to.

Buffon then quotes an account by Franz Pyrard, written in 1619:

> I have seen an elephant carry with its tusks two metal cannon tied and fastened together with ropes, each cannon weighing 3000 lb.; it raised them from the ground without assistance and carried them 500 paces. I have also seen an elephant drag seagoing craft and galleys from the water's edge on to dry land, and refloat them.

In another interesting account Buffon says:

> Once tamed, the elephant becomes the gentlest and most obedient of all creatures; it becomes devoted to its keeper, displaying great affection for him, anticipating his wishes and seeming to guess what will please him. In a short while it learns to understand signals and even words of command. It can distinguish between the tone of command and that of anger or of satisfaction, and acts accordingly. It does not make mistakes over its master's words of command, but receives them with attentiveness and executes them with intelligence, though without undue haste. It is harnessed with straps to carts, ploughs, ships, and winches, which it pulls along steadily and unremittingly, without resisting.

All the sources drawn on by Buffon, as well as those of Brehm about a hundred years later, agree that the way to obtain the best results from intelligent working elephants is to encourage them with words of appreciation. This is precisely what is done to this day by elephant keepers in Burma and trainers in every circus the world over. If, for example, an elephant is set the task of pushing with its head a fully loaded railway truck a few dozen yards to the platform, it is spurred on with words of praise. If in addition it receives a small reward, such as a bun or a few carrots, it will be all set to go on to the next strenuous task it is given.

Elephants can often be observed performing their tasks on their own initiative, without any guidance or instructions; some of the twenty-six elephants which the circus owner Carl Krone took with him on his tour through Austria, Hungary, and Italy in 1930 cleared large stones and girders to one side on their arrival at the goods depot without being told to do so.

In 1950 I saw elephants belonging to a famous trainer turn on taps and take up pails with their trunks and fill them with water, without a single word of command.

If the circus pitched camp near a field and the owner did not mind the elephants browsing there for a few hours, they would make straight for the trees and bushes to sample the foliage. About noon they would seek the shade of the trees to shelter from the sun's heat. But shortly before the 2 p.m. performance was due to begin, these intelligent creatures would make their way of their own accord to the Big Top. One of them would move the free part of the paling aside with its trunk till the gap was wide enough for them all to march through. On one occasion the paling collapsed flat on the ground between the elephants and the circus. Although it would have been easy for them to walk across it, they did not, but instead cleared the paling further over to one side, as though aware that by treading on it they would have splintered it.

As early as 1669, Father Philipp of the Order of the Blessed Trinity described in his book *Journey to the Orient* similar behaviour on the part of elephants, at a shipyard in Goa:

> A number of very heavy beams were lashed together and the end of the rope was then thrown to the elephant, which took it into its mouth and wound it twice around its trunk. Unaided and without a keeper, the elephant proceeded to drag the beams to the place where the vessel was under construction. What I admired most, however, was that when the elephant encountered other beams which hindered its progress, it placed its foot under those it was dragging and raised them at one end so that it could continue on its way. What more could the most intelligent of men have done?

Similar accounts were given by de Feyne, who in 1630 wrote about the elephants "in the realm of the Great Mogul", by La Boullaye-le-Gouz, who in 1657 reported on the training of elephants in Ceylon, and by J.-B. Tavernier, who published a book in 1679 on his travels in Turkey, Persia, and the East Indies.

In 1713 S. Thevenor wrote in his *Reisebeschreibung* (Description of a Journey) about the five hundred elephants belonging to the Great Mogul's household which were decked out in exquisite finery and behaved with a decorum in keeping with the pride they felt at being so honoured. He describes vividly the docile conduct of those elephants whose task it was to convey with gentleness and dignity the ladies of the Mogul's harem in their latticed litters: on arrival they lowered themselves to their knees with such deliberate care and elegance that "each lady was able to alight with great ease from her litter", being helped down with the unbidden aid of her mount's trunk.

In 1691 the Italian Count Aurelio published a book in Parma describing similar events observed in the kingdom of Tonkin, "where ladies of rank are accustomed to riding on elephants, which are so large and so strong that they can carry a tower containing six persons and also the driver, who sits on the elephant's neck".

Such intelligent creatures had a very high market value. S. Ribeiros, in his history of Ceylon printed in Trévoux in 1701, quotes the prices then current in Asiatic countries for well-trained elephants: "Elephants in Portuguese India are known to have been purchased for as much as 15,000 rupias. An elephant from Ceylon is worth at least 8000 pardaos, and even 12,000 or 15,000 if it is really large." In Ceylon, a family house in town at that time cost 800–1500 pardaos, and a country house in Delhi with a small park, suitable for a noble family, 12,000–15,000 rupias. The annual wage of an elephant keeper was about 300 rupias, plus free board and accommodation—in the elephant's quarters! In every country of southern Asia elephant keepers were among the best-paid skilled workers and, then as now, were treated with a great deal of respect.

Buffon assembled a good deal of material concerning work-elephants in India in the seventeenth and eighteenth centuries, which was published about 1770. He states:

> As they combine intelligence with strength, they never break or damage anything entrusted to them. They go on twisting and turning bales until they deliver them safely from the water's edge into the vessel, never allowing them to become in the least wet. They set them down gently and arrange them as required. They inspect them with their trunks to ensure that they are lying properly. If there is a barrel which

could roll, on their own incentive they find stones and place them so that it remains firmly in position.

It is clear, therefore, that over the last three hundred years quite a lot was known about the accomplishments of Asiatic elephants, and knowledge of African elephants was increasing too. Naturally great efforts were made to exhibit these clever creatures in Europe. In 1668 Pedro, the twenty-year-old Regent of Portugal, sent an African elephant which had been captured in the Congo to Louis XIV of France. It lived for thirteen years in the menagerie at Versailles, and was then transferred to Paris to the Jardin du Roi. There it was so indulged and overfed that in 1681 it died of digestive disorders and obesity. Every day it was given 80–100 lb. of bread and two buckets of bread-soup or cooked rice, as well as several buckets of water laced with a gallon or two of wine, not to mention all the corn and grass constantly available and the titbits given by visitors. With such a diet it is little wonder the poor creature soon succumbed!

Some Indian elephants housed in the menagerie in St Petersburg during the reign of Peter the Great (1682–1725), and also an Indian elephant kept at Naples, were literally fed to death. Elephants, which display such remarkable intelligence in every other way, show none in their craving for food. If instead of being put to work they are given a life of ease and plenty, their mental faculties soon deteriorate—in much the same way as with humans.

Although tame elephants have been reared for centuries in the countries of southern Asia, there are never enough to meet the demand. The stocks in Burma, Ceylon, and Thailand are still supplemented by captured wild elephants.

Their capture, just as their taming, depends on the help of domesticated elephants. Individuals from a wild herd are segregated in an enclosure by trained elephants, and then bound with chains by the elephant catchers. Then two large trained elephants make for the strongest member of the herd and edge it right up against a tree, so that its hind legs can be lashed to it. Without the intelligent co-operation of the tame elephants this task would be extremely difficult and dangerous for the elephant catchers.

As early as 1679, in a book entitled *Les six Voyages de Jean-Baptiste Tavernier*, there is a description of the assistance given by tame elephants:

> If the wild elephant did not do as it was commanded, the people ordered the tame elephants to punish it. This one of them did forthwith, striking it about its forehead and head with its trunk. If the wild elephant made to retaliate, the other tame elephant struck it from the other side, so that the poor creature scarcely knew which way to turn, and in such manner learned to obey.

The willingness of trained elephants to school wild ones was similarly described by the British zoologist Sir J. E. Tennent in 1867, and his account was quoted by Brehm, who corresponded a great deal with him, in his famous *Illustriertes Tierleben* (Illustrated Animal Life).

However tough trained elephants might be while taming wild ones, they change their attitude towards the newcomers as soon as they become completely domesticated. Trained bull elephants are often very indulgent towards younger females, and the tame females often display great consideration towards young males during their training period. In this way it is quite common for friendships to spring up between the 'teachers' and their pupils, which sometimes last for years.

The friendship formed between the newcomer and the 'teacher', following the brief period of discipline, leads the alert newcomer to realize the great advantage of not rebelling or exposing itself to corrective treatment from the human trainer's goad. Training is always carried out with two *kornacks*—the driver of the tame elephant and the man who will be taking charge of the new one. As soon as the new elephant has been sufficiently broken in by the tame one, its appointed driver approaches it in a friendly way, throwing food to it and later letting it eat out of his hand. Then he walks right beside the animal, still half wild, and strokes and pats it, speaking soft words of encouragement—till at length he is accepted and even welcomed as a friend. It says much for the trainer's skill that an elephant, once it has given a *kornack* its friendship, does not welcome any other on its back. Many elephants overwhelm their drivers with an extraordinary degree of affection, allowing them to stroke their trunks and following them around like faithful dogs. This devotion is one of the most satisfying results of training, and there is something miraculous about such a relationship between this giant among creatures and its master, intellectually superior but vastly inferior physically.

If an elephant has been properly trained for circus performances, it is easy to see by its general behaviour in the ring how much it enjoys carrying out its orders and responding to every gesture made by its trainer. It expresses its joy by high-pitched noises, and regards it as the height of bliss if its trainer goes over and strokes its trunk or pats its head and allows himself to be lifted onto its neck with its trunk.

The fact (unknown to the zoologists Buffon and Cuvier) that wild elephants, with their liking for a definite hierarchy within the herd, often choose a cautious 'lady president' as their leader was first observed, among Europeans, by Sir J. E. Tennent. This had been common knowledge for centuries to Indian elephant catchers, but because of the great social gap which usually existed between the British Raj, which governed practically the whole of India, and the indigenous population it was kept a close secret by the Indians for the benefit of the elephant catchers.

Of the very few Englishmen who learned the secret Tennent was the first, partly because of his interest in zoology and partly because he enjoyed the confidence of a large circle of Indian friends. He was able in due course to give a comprehensive picture of the daily life of the Indian elephant in a book entitled *The Wild Elephant*, published in 1867. In this he states:

> Elephants have a very great feeling for regularity and order, and every herd chooses a leader which is strong or clever enough to protect it. Occasionally a female is chosen, if she exceeds the bull elephants in reason and judgment. Whether the herd leader is a bull or a cow, once chosen it finds strict obedience in all members of the herd.

African explorers have confirmed that the herds of African elephants living in the jungles and the bush do not unreservedly trust in the strength of the strongest member, either, but often bow to the authority of a mentally alert and physically impressive female. Such 'democratic equality of the sexes' is almost unique among mammals leading a communal life.

An interesting explanation of the reason why an elephant herd is sometimes led by a female is offered by the French zoologist Georges Blond: "The instinct of preservation of the species has caused the cows to defend the bulls against humans who lie in wait for them"—in order to obtain ivory. According to Blond, the cows are anxious for the safety of the herd and attack first, since elephants certainly do not need tusks to settle accounts with man, as their trunks can be used to devastating effect. Blond is of the opinion that these female strategists and tacticians bend the males in the herd to their will and assume the status of commanding officers.

Evidence from a number of experts on elephant lore shows that African elephants living in the wild under the leadership of shrewd females have achieved some remarkable feats. No less an authority than the French writer Albert Jeannin maintains, in his book *L'Éléphant d'Afrique* (1947), that elephants can construct full-scale dams. While it would be necessary to presuppose considerable mental ability on their part to ensure for themselves a constant water supply through the use of tree-trunks, clods of earth and so on, it would be wrong, in view of what we know about trained elephants, to dismiss such reports out of hand simply because no such observations have been noted among trained Indian elephants. Creatures living under man's protection scarcely need to protect themselves against thirst. Moreover, the regions of southern Asia inhabited by elephants are much better provided for naturally with water than those inhabited by African elephants.

CHAPTER TWO

SERVICE IN WAR AND PEACE

FORTUNATELY, domesticated elephants have not been used as 'shock troops' in war for centuries now. As reserve forces, however, they have retained a certain importance from the time of Akbar up till now in Asiatic countries. They owe their transfer to base not to man's respect for them, though, but rather to tactical considerations. The fear and bewilderment which they once inspired in the enemy, and which had quite significant consequences even in medieval times, was markedly reduced with the increasing use of explosive missiles and changes in methods of combat.

A story, dating from the time when the French still maintained troops in south-east India to protect their trade interests there, illustrates the importance attached to good understanding between an elephant on war service and its driver (known in India as 'mahout').

A soldier belonging to Governor Dupleix's Elephant Corps, which formed a labour force within the framework of the military forces of the French East India Company, was stationed with his elephants in Pondicherry, a port on the Coromandel Coast, from 1730 till 1741. This mahout used to convert his pay into arrack (a spirit distilled from rum and flavoured with eastern plants and fruits, very popular in India), which he would conscientiously share with his elephant. It is hardly surprising that the creature formed a close attachment to its master, seeing that watered-down arrack is one of the favourite drinks of elephants.

One day the soldier drank more of this potent mixture than was good for him. The military police, who as a rule were very tolerant towards mahouts, found themselves obliged to take the noisy fellow into custody for the night to sober up. But he managed to escape and was ultimately found in the elephant's quarters, where he lay snoring peacefully alongside his elephant. It was impossible for the military police to approach him because the elephant warded them off by raising its trunk threateningly and trumpeting loudly. In the archives of the French East India Company it is recorded that from the time of this incident it was forbidden to punish mahouts—a decision which no doubt took into account the fact that elephants were urgently needed for heavy work, and it was impossible to make the elephants perform it without their mahouts.

The British forces in India also kept many work-elephants, as well as others which appeared on ceremonial occasions decked out in all their ostentatious trappings.

Following India's independence and the formation of the Indian Union and the Republic of Pakistan, the custom of keeping work-elephants paid for by the State was retained, and they were still also used from time to time on ceremonial occasions. In the Indian Union they are assigned to the armed services in the various hill regions as 'living tractors' for hauling cannon, and they have their own personal attendants. In every battery there is an elephant officer, to whom the twelve elephants serving as draught animals for six cannon and the mahouts are subordinate. For each elephant their is an attendant with military rank who is responsible for providing fodder and also for making daily reports on the elephant's health to the mahout. According to service regulations, these elephants, which are very well cared for, are expected to cover a maximum of 38–44 miles in a day's march.

Even the army has to accommodate itself to the elephants' special work rhythm and make concessions to them. Work-elephants know their hours of work exactly to the minute. On the basis of centuries of experience in the civilian sector, army elephants must be granted leave for rest and recreation for several months each year. The elephants in the Indian army see no reason why, simply because they belong to the armed forces, they should renounce their well-earned rights. They therefore enjoy special privileges, from which the mahouts also benefit—namely, extensive leave.

More modest marching distances were required of Jenny, an Indian elephant who, during the First World War, served in northern France as the last German emperor's 'heaviest soldier'. Her sphere of activity was approximately a hundred miles west of Aachen—the town in which an elephant named Abul Abbas worked for the first German emperor, Charlemagne, who used also to ride on it. There, in the pine forests between the villages of Flaumont and Felleries, about six miles from the Belgian frontier, it was Jenny's task to haul felled trees to the sawmill. She also occasionally helped to shunt railway trucks loaded with planks and timber destined for the trenches—a job she had learned in peace-time in Carl Hagenbeck's circus and in the Hagenbeck animal park at Stellingen. This willing and able creature belonged nominally to the army commanded by Crown Prince Wilhelm, though in point of fact she showed no signs of obedience to this much-decorated general, but only to a plain bosun's mate, who in peace-time had been her keeper in the Hagenbeck animal park.

The bosun's mate was frequently involved in amusing incidents with his elephant. High-ranking officers who came to inspect the troops were amazed to find this unlikely pair on the parade ground. In the Army Manual no provision had ever been made for bosun's mates with elephant goads. Because of his outlandish uniform he was sometimes mistaken for a mad gipsy.

Lorenz Hagenbeck has paid tribute to this remarkable female auxiliary in his book *Den Tieren gehört mein Herz*: "Some evenings she would give a special performance of her circus tricks, and the laughter of the troops, seated on the felled tree-trunks around the improvised circus ring, would echo through the forest of Felleries."

When the troops were forced to withdraw Jenny was taken

Alpine rescue dogs at work after an avalanche

112

Lumpi makes himself useful *Photos: Conti Press*

Not every Alsatian can perform like this *Photo: Dr Jesse*

back safe and sound to Hamburg-Stellingen, where she was given a period of leave after all her exertions. She later returned to circus life.

During the Second World War, too, elephants played an important part militarily, though not in France as Jenny had done. But they were used on a large scale in the war in Burma.

The elephants used in that war formed part of the six thousand, all fully trained and capable of working on their own initiative, belonging to the Bombay-Burma Corporation, which were scattered over a number of stations in the forests of Burma. Their task was to transport teak to the depots. 'Specialists' among them were put to work in the sawmills, where they displayed considerable team spirit. They were employed as fetchers and carriers, besides being responsible for stacking the sawn timbers. Selected bulls then carried the teak from the mills to the quayside, making use of both their tusks and their trunks. The cows were unable to perform this task, as they have no tusks, but they made themselves useful in other ways. In fact, without them the whole teak exploitation programme would have been impossible, as the bulls were comparatively few.

In 1942 the war spilled over into Burma when the Japs invaded it from Thailand. The situation changed dramatically overnight for the managers of the work stations, the forestry officials, and also for the *oozies* (drivers) and their charges. The Bombay-Burma Corporation, though a private undertaking, immediately placed itself under the command of the State and the Army. The orderly withdrawal of the English civilian population in the face of the advancing Japs was largely due to the unrelenting devotion to duty of the elephants.

The elephants were put to work part of the time on the building of roads and bridges. In this they showed how cautious and distrustful they are: they dislike ground which is not firm and often refuse to walk up a gangway onto a ship, which is why they are usually placed in a special harness and hoisted on board by crane or derrick. Elephants which are not yet fully domesticated are shipped from point to point in solidly constructed crates.

In the Burma campaign it was impossible to induce the

elephants to walk across bridges which, built exclusively for pedestrians, seemed to them too frail to bear their weight, yet they were quite willing, after a few tentative steps, to cross bridges which they themselves had helped to build for the use of army vehicles: it seemed as if they were able to gauge the weight such bridges would bear.

Many elephants and their keepers were later captured and forced to work for the Japs. Thousands of *oozies* managed to escape with their beasts into the jungles of the north, however, where they kept themselves hidden until the British had driven the Japs out.

Only forty-five keepers managed to escape with their elephants across the mountains into Assam. These returned with the 14th Army to Burma along the roads which by then had been made safe for them. With the help of elephants it was also possible to bring food supplies into hunger-stricken mountain areas impassable to jeeps. Because of the exceptionally hazardous condition of the roads in Burma elephants have played an important part since the Second World War in transporting passengers and goods.

To mount their elephants more easily, and also to spare them the difficult manoeuvre of standing up with heavy loads on their backs, the keepers use a method devised over a hundred years ago: the elephants are trained to raise their right or left foreleg and to hold their trunks out sideways, so as to provide two steps up for their keepers. Some are trained to go one step further by raising their keepers gently into the air with their trunks and setting them down comfortably on their necks.

Asiatic elephants have become rare in their old homeland, India. The number of those living in freedom has diminished over vast areas so drastically—because of the population explosion, large-scale rounding up and the earlier massive slaughter of the species for ivory—that for hundreds of miles round such cities as Benares, Bombay, Delhi, and Calcutta there is not a single wild elephant to be seen. Stocks of domesticated elephants have become considerably smaller, and the demand for work-elephants has disappeared. Since maharajas now have to pay an elephant tax, far fewer are kept for ceremonial purposes.

Elephants as a means of transport have been superseded by Bentleys, Mercedes and Rolls Royces.

Nowadays very few people own elephants. Among those who do are authorities in charge of important temples and holy places, and a handful of wealthy lay dignitaries.

From 1947 till the mid 1950's the majority of the elephants which belonged to the previously independent ruling princes of India were purchased by large animal dealers in Europe and the United States of America and by numerous zoos all over the world. Almost all the maharajas of the former native states offered their elephants—"guaranteed tame, well-trained and safe on the roads"—for sale, and the sale catalogues described at length their acquired talents and abilities.

When an elephant was sold to an overseas customer an 'instruction handbook' was supplied with it, so that the new keepers would know what to do with their acquisitions. Their former owners, the maharajas, deserve a special tribute for their continued interest in and affection for their old friends, as even after they had been sold inquiries were made as to their state of health and how they were settling down in their new surroundings. Often details which had been overlooked about their particular qualities were sent on. It is clear that the princes were reluctant to part with their elephants. Even today elephants belonging to Indians are offered for sale "at a cheap price, owing to circumstances". Whether the price really is cheap, however, is questionable, for by the time the foreign purchaser has settled the freight bill and paid for food supplies and an experienced attendant for the voyage, he will have incurred considerable costs over and above the 'cheap' purchase price. But wherever the maharajas' old friends may be now, scattered all over the world in circuses and zoos, their new owners have nothing but praise for them and their intelligence, their readiness to learn, and their docile behaviour.

Nowadays elephants can be obtained simply by placing an order by telephone. If a circus proprietor in South Africa or North America wants to purchase a group of well-trained, intelligent Indian elephants, or the director of Ueno Zoo in Tokyo wants an African bush elephant to extend the range of his elephant house, the interested parties just contact the firm of Hagenbeck in Hamburg-Stellingen or in Alfeld. A significant part of the world trade in large animals is carried out through these channels.

The German trade in large animals reflects a whole century of tradition. As far back as the 1860's Carl Hagenbeck sent out his own animal collectors to Africa to bring back elephants and other wild animals for him.

In 1866 the first large consignment arrived in Hamburg from the Sudan by way of Trieste, Vienna, and Nuremberg. Others soon followed. One of these consignments, consisting of elephants, giraffes, ostriches, and other creatures, was assembled by Hagenbeck to form a 'Nubian caravan', which attracted hundreds of thousands of visitors to the Berlin Zoo, even though at that time it was situated well outside the city in a sparsely populated sector.

Where the Kaiser Wilhelm memorial church now stands there was then a stop for horse-drawn omnibuses, which jogged their way in half an hour to the Potsdamer Platz and in forty-five minutes to the Alexander Platz. Most people went to the Zoo by horse-drawn cab—wealthy people in a landau or their own coach. One day a certain Professor Virchow paid a visit, to see the African collection which had become the talk of the town. The Professor, an authority on wild life, told Carl Hagenbeck that in his view African elephants, unlike Indian elephants, were untrainable. Hagenbeck, however, who had already had a number of years' experience with African elephants, was convinced that they were capable of being trained. He offered to train the five newly arrived African elephants to the point where, within twenty-four hours, they would be willing to take a man for a ride on their backs. Professor Virchow regarded this as quite impossible.

As soon as the celebrated professor had left the Zoo, Hagenbeck arranged for some of the Nubian attendants who had travelled the whole distance with the caravan to mount the elephants and, despite some initial resistance, to ride on their backs. Cajoling them with buns, succulent green vegetables and carrots, the Nubians won over three of the elephants by evening, and next morning the other two followed suit. "It now only remained to persuade them to carry loads instead of men", Hagenbeck recalls. "I ordered some sacks to be filled and bound together in pairs with straps, and these were then hung over the backs of the elephants. The beasts at first disliked the feeling of the loads resting against their flanks, but they soon became accustomed to this sensation also."

When Professor Virchow arrived at the Zoo at the appointed hour Hagenbeck was able to demonstrate to him that African elephants can indeed be trained. The Nubians rode the docile beasts up to the distinguished guest and then rode them away with loads on their backs.

That summer's day in the mid 1870's is in fact one of historical importance: since Hannibal's march across the Rhône and the Alps with elephants the recorded instances of the Romans attempting to domesticate African elephants are very few, and as time passed knowledge of the teachability of this species was lost—so much so that it was taken for granted that it was pointless to attempt to train them. Carl Hagenbeck's hasty training programme showed that it was indeed possible, and this contributed to the establishment some years later of a Belgian training centre in the Congo. This centre, set up at the end of the nineteenth century, was at first situated at Api, near the present-day Congo elephant reserve. It was later transferred to Aru, a border town which separates the Congolese Republic from Uganda, about 75 miles north of Lake Albert.

There are many examples which show how much the tame, docile elephants here responded to kind treatment by their human trainers. One lame bull elephant was taken a long way from the domestication centre, where it had spent most of its life, and set free; but instead of joining up with the elephants that roamed wild in its new surroundings, it made its way straight back to Aru. Another attempt was made to give it its freedom, but that failed too. It always found its way back and hung around the entrance until it was brought in, refusing every chance of liberty in favour of human company, prepared meals and familiar surroundings. It was allowed in the end to stay on permanently at the centre, although it limped, as an 'instructor' of young captured elephants.

The Belgians had a similar experience at Aru with a number

of completely domesticated cow elephants. They had escaped and could not be recovered, despite an intensive search. Yet a few days after the search was abandoned they arrived back at the entrance, trumpeting to be let in, as though their brief flight to freedom had been a bitter disappointment.

The *Station de Domestication des Éléphants* still exists and is now maintained by the central government of the Congolese Republic. The Belgians working at Aru during the disturbances in the Congo between 1961 and 1965 were not molested by the native population, and today fully trained elephants are still transported from there to America and Europe, to find their ultimate home in a circus or a zoo. Some of these elephants are further trained on arrival for performances in the circus ring.

Carl Hagenbeck was an eyewitness to the violent reactions of his highly intelligent elephants to fire, smoke, and loud noises.

On 31st July, 1888, to mark the centenary of his family's business enterprises, Carl Hagenbeck had taken his circus to Munich. Eight elephants had been selected to lead the procession. Four of these were ridden by experienced keepers, the other four by men who knew the creatures well. The noise of the crowds lining the streets made little impact on the elephants, who were used to loud applause in the circus ring and to the clatter of the circus being erected and dismantled. Spectators threw apples and buns to them without their becoming in any way disconcerted, though they only took those which fell immediately in front of them.

In the Ludwigstrasse the Prince Regent took the salute, with the elephants raising their trunks as a mark of honour. On the return journey they had to pass a steam-engine adorned with painted cardboard so that it resembled a fire-breathing dragon. One of Hagenbeck's officials had expressly told the driver and passengers to stop the engine, shut off the steam, and refrain from using any whistles while the elephants were going past. But the driver and passengers, who had been celebrating the occasion with good old Munich beer, ignored the instructions—no doubt because they thought it would add to the fun, as they could hardly have been aware of the disastrous consequences their high-spirited behaviour would have. And so the driver let off steam and blew his whistle just as the eight elephants were alongside his engine. The elephants were already apprehensive of this monster the like of which they had never seen before, and four of them bolted across the street and down through the Odeonsplatz. Panic quickly seized the crowds. Hagenbeck was at first able to control the other four elephants, but then a gang of tipsy rowdies advanced towards them with sticks and knives and struck at them, jabbing them in the trunk. Hagenbeck himself was pushed out of the way by the howling mob and nearly trampled underfoot, and the terrified elephants stampeded.

Mounted troops were called in to head the elephants off, but their gleaming swords only made things worse, and it was not until the keepers finally managed to reach them that the elephants were pacified.

Because of the engine-driver's thoughtless action a large number of people were injured, some of them by being knocked down by the militia's horses. Yet, extraordinary as it may seem, not a single person was harmed by the stampeding elephants.

In every circus the largest elephants are enlisted to help put up or take down the tents when there is a shortage of labour. If in wet weather the tractors are unable to tow the heavily loaded trucks over the soft ground, the elephants are given the job. Their amazing achievements have certainly contributed to the high prestige which German circus owners enjoy on their foreign tours.

During their tour of Austria, Hungary, and Italy in 1930 the Krone Circus often had to cope with prolonged periods of rain, especially on the days of departure. When the circus props had to be taken down overnight, the task was conscientiously carried out by the largest of the circus's twenty-six elephants. Although they had only just finished doing their tricks at the last performance, they willingly allowed themselves to be harnessed and in pouring rain set about hauling the heavy trucks across the soggy ground to the made-up roads. They never refused to work, even in the worst weather. Carl Krone gained the impression that the elephants realized that their help was being depended on.

After Lorenz Hagenbeck had delighted millions of spectators

in various parts of Japan throughout the summer of 1933 with his trained elephants and other performing animals, his whole Noah's Ark was shipped off from Osaka, the last Japanese town of the tour, to China, in a specially equipped German steamer. Pouring rain had completely softened the ground on which the circus was to be erected in the centre of Shanghai. The circus trucks got stuck in the mud up to their axles. It was only thanks to the elephants that the opening performance could begin on time. They did not mind the strenuous work in the rain at all—in fact, they revelled in it, treating themselves to a mud bath, which the wild elephants of Assam, Burma, and Ceylon are so fond of. Hagenbeck comments: "But when our working elephants plunged to their hearts' content into the mire, trumpeting with delight as they wallowed in it, everyone began to feel happy again." Their help was again needed when, because of the streams of visitors to every performance, the tent was in danger of collapsing: the willing beasts stood, unflappable, like a living wall, supporting the front of the main entrance.

"If animals could be awarded medals", Hans Stosch once said to me, after telling me about the way elephants had helped to erect and dismantle circus equipment, "the first to be honoured would be our elephants, which have such keen judgment and are so civilized!"

On countless occasions I have watched elephants in their role as transport workers, pushing heavy circus trucks with their foreheads (protected with leather padding) or in harness hauling their loads, working with precision and never jibbing. The most remarkable thing is that when they enter the circus ring, even just after a really heavy stint, these good-natured animals, with their artistry, are still the stars of the show.

The efficiency with which elephants work, though taken for granted by many people, does in fact presuppose a high degree of mental ability. The precision shown by Hagenbeck's elephants from Hamburg-Stellingen particularly impressed the director of Rome Zoo, Dr Knotterus-Meyer, "when one of them pulls eight to ten loaded lorries behind it as though it were going for a stroll. Or when, at a signal from its keeper, it kneels down to move large lumps of rock with its forehead, then gets up to re-position itself before repeating the action over and over again till the rock is in the right place. Or when it carries girders or tree-trunks, after establishing exactly where to hold them to distribute the weight evenly."

About his own experiences in training elephants he says: "The usual tricks, like playing a barrel-organ, dancing, passing coins to their keepers, which they quickly learn in return for a bun or some fruit, expressing thanks and making requests by going down on one knee and trumpeting—such things elephants have no difficulty in learning."

Further evidence of the intelligence of trained elephants was provided in Hagenbeck's animal park by a female named Ruma. Without any prompting she took to disciplining the less 'civilized' of her companions. The younger ones were roped together with her, so that she could take them on a tour of the park some mornings when there were not many visitors, to get them used to human company. If one of her pupils seemed at all reluctant, Ruma asserted her authority by striking it hard with her trunk. Her corrective treatment never failed to achieve the desired results. The elephants she had instructed became in time as gentle as lambs.

CHAPTER THREE

GOOD CHARACTERS

THERE are any number of stories to illustrate what good memories elephants have, their gratitude for kindness shown to them, and their tendency to bear grudges, sometimes for years, against people who have offended them in some way.

Plutarch tells of an elephant which took revenge on its keeper, who had mixed stones and earth in with its feed, by seasoning his bowl of food with sand.

According to one seventeenth-century account a tame elephant in southern India recognized a soldier, who a few days before had blocked its path and refused to budge, sitting on a river bank with some friends. The elephant got its own back by seizing the soldier round the waist with its trunk, lifting him over the heads of his friends and dumping him in the river. It then lifted him out again before going calmly on its way. The behaviour of this elephant, as described here by Buffon, suggests a measure of revenge tempered, however, with moderation, as if to warn the offender not to take any further liberties.

All domesticated elephants—even the most intelligent—have a weakness for alcohol. For centuries the Indians have exploited this: if they want their elephants to carry out work of a specially arduous kind they encourage them by giving them a taste of millet beer or arrack punch, and the elephants then know that once the work has been done they will be given a full quota. But it is essential for the keeper to keep faith with them, as otherwise they would almost certainly turn on him and show him no mercy.

In the hilly region of Bangalore, in the Deccan, one elephant took its revenge gruesomely on a keeper who had cheated it of its promised reward for services rendered by stamping on his head. The victim's widow, in her grief, offered herself and her children to the enraged beast, exclaiming that she did not want herself or her sons to go on living now that her husband had been killed. The elephant calmed down immediately, as though regretting what it had done. Then it picked up the bigger of the two children and placed him on its neck, to show that it wanted him as its new keeper, and after this it would never allow anyone else to give it orders.

Anyone acquainted with the ways of domesticated elephants will not doubt the truth of this story as told by Buffon. It is quite probable that the keeper who had failed to keep his part of the bargain had already allowed his elder son to ride the elephant and had given him lessons in how to manage it. It is still quite common to see boys in their early teens riding elephants and giving them instructions. As all Asians treat elephants with respect, and children are instinctively kind to them, it is natural that most elephants prefer a young lad as their driver.

Visitors to the Jardin des Plantes in Paris in 1798 witnessed an amusing incident involving one of the two elephants which had been transferred there from the Netherlands. In front of the sturdy bars of their enclosure a sentry stood guard, because both elephants had been captured in the war between the two countries. Knowing how visitors like to feed the animals, the director had put up a notice forbidding the elephants to be fed, and it was the sentry's duty to ensure that this order was observed. Most of the sentries turned a blind eye, but one took his duties very seriously and prevented the public from giving the elephants anything at all to eat.

Parkie, the female, who was more intelligent than her male companion (as is often the case with elephants), quickly realized how unwelcome the presence of this particular sentry was. She waited till he was just passing her, then seized his rifle with her trunk, pulled it through the bars and with one foot bent the barrel almost at right angles. She then tossed the weapon back at him and, for good measure, squirted water at him from close range. For a day or two afterwards Parkie was the talk of Paris.

At the turn of the nineteenth century there was quite a sensation when a tame female elephant, about twenty years old, was unloaded at the East London Docks from a clipper which had brought her from India and was taken to the Duke of Devonshire's residence in Chiswick. A specially heated building had been erected for her in the grounds, and the butler had been briefed on how to handle and feed the distinguished guest.

From the moment she arrived the young lady elephant displayed a warm affection both for the Duke and his butler. She seemed to sense how exceptionally fortunate she was. In India she had been trained to do various kinds of work, and now in Chiswick she seemed keen not to lead an idle life. And so, among other tasks, she carried a huge bucket, usually managed with difficulty by two strapping men, for the butler, who was also in charge of the estate. And, after being shown just once, she pulled the water-wagon along the paths of the estate, and helped to clear surplus trees, simply by leaning heavily against them. At a given signal from the butler she uprooted bushes and even learned how to sweep up leaves and twigs with a besom, which she manipulated skilfully with her trunk.

When the Duke had guests their children were allowed to climb onto the seats strapped to her flanks and be taken for a ride through the park. Previously, in India, she had been given the care of small children and had rocked babies in their cradles. Now, by the Thames, she showed her liking for children again. She let them lead her through the park and looked on them as her protégés. For her services she was generously rewarded by the older children with fruit, buns, and carrots.

The Duke enjoyed telling his guests how she could open unaided a bottle of fruit juice or watered-down whisky, provided the cork was not too tightly fixed. To do this she laid the bottle in a sloping position on soft ground, held it firm with one foot and drew the cork with the tip of her trunk. She then grasped the bottle neatly in her trunk and poured the contents into her mouth. The Duke was always prepared to lay bets with his guests that she would perform this trick without spilling a single drop. As if sensing that both the Duke's honour and her own were at stake, she always succeeded in winning the bets for her owner.

Over a period of years Dr Knottnerus-Meyer made a detailed study of a tame African elephant named Toto. Whenever he appeared with a piece of chocolate Toto would come hurrying along; but if he appeared with nothing the disappointed elephant would spray him with sand. Before allowing his keeper to remove his empty rice bowl, Toto would first insist on making quite sure that there was nothing left in it. If the keeper neglected to show him the bowl before going off with it, Toto would trumpet with annoyance until he did so.

Some of Dr Knottnerus-Meyer's observations about Toto are paralleled by others I made myself some years ago concerning a female Indian elephant belonging to the Krone Circus. Assam was one of the nicest and most docile creatures you could wish to meet. One day she removed the straw hat from the head of a newspaper reporter accompanying me round the elephant house and ate every scrap of it. Dr Knottnerus-Meyer reports: "One thing which gave Toto great pleasure was to appropriate ladies' hats through the bars of his cage, which he did with the speed of a monkey."

Though they daily consume two or three stale loaves and a large quantity of oats, bran, or coarse-ground corn as well as a sizeable pile of freshly cut grass or a less voluminous quantity of turnips or other root-crops, elephants still enjoy an extra ration of hay or straw, so it is not surprising that an occasional straw hat is welcomed as dessert. The director of the Leipzig Zoo has observed that elephants are equally partial to wickerwork baskets and besoms, and have even been known to eat leather briefcases. One elephant made a meal of a rucksack!

Toto demonstrated that he was just as skilful as any Indian elephant at turning on hydrant taps. Walking through the streets of Dar-es-Salaam one day advertising the circus like a well-trained sandwich-man, he manipulated the lever of a water-pump with his trunk and quenched his thirst.

At the Rome Zoo Toto made friends with an Indian female named Greti. One day, having injured his trunk, Toto was confined to his enclosure. To avoid hurting his trunk, his food was placed direct into his mouth and he was given no hay. Greti made up for this by taking across to him from her adjoining enclosure several trunk-loads of hay from her own pile and pushing them through the bars. Dr Knottnerus-Meyer comments: "I was dumbfounded by this Good Samaritan gesture and must confess that I would have found it difficult to believe if I had not seen it with my own eyes."

CHAPTER FOUR

PRACTICAL HABITS

It is not at all difficult to illustrate the prudence and discrimination shown by elephants even in their ordinary daily routine. For instance, when being transferred from railway trucks to road trucks, they prefer to step straight from one into the other, without a ramp being used. If this is unavoidable, however, they carefully test the ramp to make sure that it is firm enough to take their weight.

Large elephants leave trucks rear first if the doors are lower than shoulder height. They do this by taking up a half-kneeling posture as they squeeze out, snorting and groaning.

When elephants first arrive at zoos it is remarkable how cautiously they inspect the artificial ponds provided for them in their enclosures. They kneel down at the water's edge and gauge the depth of the pond with their trunks. If satisfied they enter it gingerly, probing the steps leading down with their trunks, then they tread circumspectly about until the whole bottom has been thoroughly tested. Only then, after they have satisfied themselves that the pond holds no unpleasant surprises for them, do they abandon themselves to the pleasures of bathing in it.

A couple of elephants, named Dicky and Mary, were successfully reared by the father of Adolf Althoff, a famous circus owner and animal trainer. As small boys, Adolf found his brother watched their progress with passionate interest and soon gained the affection of both of them, as the children indulged them with little acts of kindness and attention. They brought them turnips, cabbage, loaves of bread, apples, and greenstuff—and were amazed to see them trample on them to make them a convenient size for eating. Dicky would place the cabbage or bread in front of his left forefoot and then tread firmly on it until it was broken down into small pieces. This showed that he was 'left-footed', whereas Mary mostly used her right foot.

On Mary's twentieth birthday Adolf's father baked a cake-mix in the shape of a basket, coated it with chocolate, and filled it with lots of ripe yellow apples. Mary enjoyed the fruit, and allowed Dicky to share it with her. She then discovered that the container was also edible, proceeded to crush it gently with her forefoot, and offered a piece to Dicky. After tasting it herself, she trumpeted shrilly and turned her head away. Curiously, Mary did not like chocolate, but Dicky did and he had the whole basket to himself. There were other things which Dicky, in turn, did not like. Once he was offered some heads of celery and a bunch of parsley. He ate the celery, but spurned the parsley. Afterwards he drank three buckets of water one after the other, so thirsty had the celery made him, and he never again accepted celery.

Close observation of circus elephants has made it possible to assess the adaptability and resourcefulness of elephants living in their natural surroundings. Dr Hediger, for example, has shown how readily circus elephants grow accustomed to all the different kinds of noises around them. They settle down to sleep even when men are going past just a short distance away, while tractors rumble by or chains are rattling.

During my travels with circuses, which extended over ten or more years, I have often observed elephants making pillows out of hay and straw and settling down to sleep on them, quite unperturbed by my presence. A cow elephant always stood sentry if the trusted keeper was not present, but if he was they considered it unnecessary to post a sentry, unlike wild elephants, which always do so at night.

When a monkey wants to scratch itself it can reach any part of its anatomy with either its hands or its feet. An elephant, on

Photo: Professor Dr B. Grzimek

Karl Krall with his 'reading' and 'counting' Arab stallion Zarif

128

Wilhelm von Osten with Clever Hans

Photo: Historisches Bildarchiv Handke

Photo: laenderpress

Horses in training must have quick understanding and a good memory

the other hand, must resort to various expedients when it itches in one of the remoter parts of its massive bulk. Even its trunk cannot reach its hindquarters, as it cannot turn its head round as far as cats and dogs can. In the wild, elephants can rub themselves against a tree or break off a convenient branch to do the job. In some zoos, for example Leipzig, stone pillars have been specially erected for the purpose. Four thousand years ago the work-elephants along the Indus scratched their backs on the corners of houses, as can be seen from the ruins of Mohenjo-Daro and Harappa. There are no such facilities for circus elephants, and so they have the job done for them by their keeper with a hard brush, to their great delight. Often they take the brush from the keeper and use it as an artificial extension to their trunk to reach the less accessible parts of their anatomy.

Dr Grzimek has watched a trained elephant sit down on its hindquarters with a stick with a protruding nail "to scratch its belly with the nail"! "One day", he reports, "Moni took the stick out of my hand and scratched herself with it on the shoulder."

These observations, together with my own and others made many decades ago by a great variety of elephant trainers and directors of zoos, clearly confirm that elephants do use implements to scratch themselves with, in spite of some assertions to the contrary.

PART THREE
THE BIG CATS

CHAPTER ONE

LIKEABLE LIONS

"THE lion is not a vicious creature at heart," declares Carl Hagenbeck, a world authority on wild animals. He refers to them as "straightforward and sincere", emphasizes their devotion to humans to whom they have become attached, and praises their intelligence. They do not regard man as prey more or less difficult to capture. Experience and circumstances teach some to respect man as their master, while others look on man as a good friend against whom teeth and claws must never be used.

In fact the behaviour of lions, and also of tigers, suggests deliberate thought patterns. Many of these large beasts of prey are very eager to learn and adapt themselves easily to a given situation.

Observed in their natural surroundings, these two fundamentally different species at times act with quite obvious premeditation. Lions form hunting parties so that the prey can be hunted down as quickly as possible with a minimum of risk. Tigers, which as a rule 'walk alone', sometimes band together when by doing so their chances of making a kill are increased.

We will turn our attention first to the lion—the most affable of the big cats and universally acclaimed as lord of the jungle.

Aelian, Pliny and other ancient Roman historians credited lions with an extraordinary degree of friendliness—no doubt because the only lions they ever saw were tame ones belonging to public entertainers, priests and imperial rulers. Pliny makes various references in his *Natural History* to the ways in which lions were trained. Though his accounts are a curious mixture of fact and fiction, it is clear that the Romans knew a good deal about these animals. As more and more legends grew up about them, they became invested with human qualities and were credited with a degree of intelligence.

Pliny tells, for instance, what he knows from hearsay about the North African lion and then recounts a story told him by an African woman, perhaps a member of one of the Berber tribes:

> A slave woman from Gaetulia [now Tunisia and Algeria] assured me that in the forests she had soothed the ferocity of a herd of lions with sweet words and gentle speech. In order to escape a lion's fury, she dared to say that she was a female, a fugitive, a weakling, a suppliant to the most generous of all the animals, the lord of all the rest, a booty unworthy of his glory.

Stories about lions were popular with many classical Roman authors. The best known is probably the one told by Aulus Gellius and Aelian about Androcles. He was a runaway slave who, when recaptured, was sentenced to fight with a lion in the arena. While at liberty in the desert Androcles had hidden in a cave, in which he found a lion. Instead of attacking him, the lion had presented to him a swollen paw, from which he had extracted a thorn. He and the lion then lived together in a hidden cave for three years, during which time it had provided him with fresh meat almost every day. Growing tired of his solitary life, however, Androcles finally returned to his own kind, but fell into the hands of a patrol that was out searching for runaway slaves and was taken to Rome to be punished.

The lion in the arena chanced to be the same lion that Androcles had helped, and instead of attacking him it showed great affection and gratitude.

When Androcles joyfully embraced the lion and it followed him round the arena like a faithful dog, the magistrate demanded to know the meaning of its extraordinary behaviour. Androcles told his story to everyone present, whereupon both he and the lion were set free. The story goes that Androcles then went from place to place accompanied by his lion, telling his remarkable tale in exchange for food and drink.

Pliny, writing much earlier, recounts a number of similar legends: "A picture at Syracuse is evidence of this occurrence", he declares, introducing his lively account of the Syracusan Mentor who met a lion while travelling through Syria that "rolled on the ground in suppliant wise and struck such terror into him that he was running away, when the lion stood in his way wherever he turned, and licked his footsteps, as if fawning on him; he noticed a swelling and a wound in its foot, and by pulling out a thorn set the creature free from torment". According to the story, the lion subsequently visited the Mentor from time to time.

Another of Pliny's tales, richly embroidered with the author's colourful imagination, concerns a sailor from Samos named Elpis who, on landing from a ship in Africa, saw near the coast "a lion opening its jaws in a threatening way". He nimbly climbed up a tree, and the lion "lying down by the tree began to beg for compassion with the gaping jaws by which it had scared the man". The sailor called on the gods above for help. When nothing happened, he fixed his attention on the lion below, and it struck him that far from looking dangerous or bloodthirsty, the beast appeared to be desperately weak with hunger. Then he noticed that a large bone was lodged in its mouth, preventing it from swallowing food. Plucking up courage, Elpis climbed down from the tree and removed the bone from the lion's jaws. At this point in the story Pliny humanizes the lion by declaring that it showed its gratitude by bringing its catches to its benefactor as long as the vessel remained on the coast.

Provided they did not go marauding near farmsteads or vineyards in Greece or on grazing lands in Asia Minor, lions were treated in classical times with a great deal of tolerance. The townsfolk were familiar with tame and docile lions from performances given by lion-tamers and from the 'temple lions' belonging to the priests.

The Egyptian high priests enhanced their prestige by exploiting the fact that lions could be trained to perform certain functions. Lions were much sought after as pets by a number of Pharaohs because they inspired respect among their simple-minded subjects. For at least a thousand years fully domesticated lions formed part of the Pharaohs' retinues: both Ramses the Great and his queen, Berenice, for example, possessed some.

In Asia Minor priests of the Cybele cult had evidently discovered how to tame and train young lions, in order to make use of them as symbols of the power and the glory of Cybele, the "Great Mother". When the Cybele cult, which for a time established itself as the official State religion, spread across Greece into the Roman Empire in the second century B.C., it was common to see temple guardians and priestesses riding in coaches drawn by lions in the temple grounds of Rome.

Many of the rulers of imperial Rome realized the importance of these splendid creatures as a means of enhancing their prestige among the *plebs*. They employed trainers to make the lions—and less commonly tigers—so accustomed to human company that the emperors could walk about with them at no risk to themselves. Some of them even kept lions or tigers as personal bodyguards in their private apartments, while others had themselves initiated into the secret taming methods so that they could themselves harness the animals to their coaches. In fact, the lions became so tame and so accustomed to human company that they were of little real value as sentries to keep out unwelcome intruders. On the contrary, more often than not they would approach anyone in sight and expect to be made a fuss of. Unlike watchdogs, the lions were quite unable to discriminate between friend and foe and were quite oblivious of the range of human emotions expressed by speech. The four tame tigers belonging to the Emperor Claudius did not prevent his second wife, Agrippina, in A.D. 54, from handing him the poison cup, because their sense of allegiance towards her was as strong as towards her husband.

The Emperor Caracalla even took his lion with him when he set out on his campaign against the Parthians. The conspirators who murdered him in his palace in A.D. 217 were so well known to the lion who was supposed to be protecting him that they had no difficulty in overpowering the harmless creature before stabbing the Emperor to death. He had overlooked the fact that a lion cannot be trained in the same way as a mastiff.

A thousand years after the sack of Rome, by which time Nero's teams of lions, trained to draw his ceremonial coach, and Emperor Elagabalus' four tigers pulling his racing chariot were quaint stories of historical interest, domesticated lions and tigers were no longer regarded as quite so 'presentable at court'. Princes of the Italian Renaissance only kept them in their zoos. In 1573 Ivan the Terrible received a number of tame lions as a gift from Queen Elizabeth I. Not long afterwards Prague saw its first lions: the Habsburg Emperor Rudolf II appointed an expert lion-tamer from Bavaria to train the 'royal lions' at the Hradschin Palace to the point where they would be tame enough to take their places beside him in the audience chamber without attacking anyone. When this trainer died his enterprising widow took over the job, to the complete satisfaction of the Emperor and, so far as is known, of the lions too, since the chronicle does not mention any unfortunate happenings during her tenure of office as Chief Inspector of the Imperial Menagerie.

There are many other famous people throughout history who have kept lions, and most accounts agree that, as pets, they can be as loyal and affectionate as any other creature.

It is true that people used to have strange ideas about the innate gentleness of these animals. The lion-tamers kept their training methods, which for the most part relied on brutal intimidation, a closely guarded secret. Thus it is easy to see why Konrad Gesner (1516–65), under the influence of Aristotle and Pliny, could state in his book on animals published in 1563:

> So gentle and mild is the lion that he harms no creature which humbles itself before him, nor ever inflicts greater harm than is inflicted on him, desiring no further revenge or punishment to be meted out; harms man not at all, unless extreme hunger drive him to do so, much less woman than man, and children he spares altogether, so mild and just a creature he is.

It was not until some two hundred years after Gesner's book appeared that, as a result of expeditions into the African coastal regions and accounts by Arab merchants trading with Africa, a clearer picture was gained of the attitude of lions towards man. It is true that these accounts are often second-hand and wildly exaggerated, but they come closer to the truth than the earliest accounts of polar bears and tigers.

CHAPTER TWO

OBEDIENT SERVANTS

THE steady increase in our knowledge of the behaviour patterns of large beasts of prey in relation to man can be traced to a number of different factors.

Towards the end of the eighteenth century the populations of most middle and west European countries could see for themselves more or less tame lions and tigers behind bars in exhibitions promoted by itinerant showmen. 'Feeding the wild animals' became quite an attraction, which people were prepared to pay to watch, both in the travelling menageries and the public zoos in Paris and Vienna. Lions and tigers were also exhibited at fairs, including those at Frankfurt, Leipzig, Saint-Germain near Paris, Augsburg, Munich, Prague and Vienna.

In the last thirty years of the eighteenth century the first circuses came into being. Daredevil 'wild-animal trainers' began training lions and tigers to perform tricks. Most of them, it is true, were rough-necked individuals who achieved their results by means of pitchforks, whip-lashes and clubs, but from their ranks emerged in time men with a true insight into the latent abilities of these creatures.

In the nineteenth century there were some trainers who were letting it be known that, in spite of the physical problems of training in the confined space of the cages, they certainly did not look on their lions and tigers merely as 'wild beasts'—a mistaken view then still commonly held. Though few in number, these enlightened trainers showed what they could achieve with the 'big cats' when they came to be accepted not as tormentors but as real friends to be trusted.

The great break-through, which revealed beyond all doubt the innate qualities and capabilities of these animals without the use of force or brutal training methods, was made by two brothers, Carl and Wilhelm Hagenbeck, just under a century ago. They deserve their reputation as the benefactors of the 'dangerous wild beasts', for they were the initiators of modern training methods. Just about every 'big cat' trainer in the world nowadays follows their principles: to begin with, the lions, leopards, tigers, etc., are made tame by constant contact with humans, for instance at feeding time, to the point where they trust as well as tolerate them. The next stage is carried out in the big circular cage, where they are taught to accustom themselves to their own particular place in it, and after this they enter on a further stage of training, and so on.

Just how far big cats can be trained need not concern us for the moment: every circus performance involving lions and tigers demonstrates how far their intelligence can be aroused and exploited. Sometimes their masterly performances lead the impartial observer to believe that they are more adaptable than some people are! This, as every expert knows, in no way detracts from the achievement of the trainer: it just goes to show that the animals have been fortunate enough to be trained by someone of outstanding competence. When this is the case the animals come to regard training as a pleasant change from their normal routine.

One animal trainer in England takes his pet tiger for a walk every day. When he goes to the hairdresser's the tiger sits at his feet quite unconcerned while the hairdresser snips away at its master's hair. Trainers in America and Europe have been photographed on innumerable occasions taking their pet lions into sweet-shops or their pet tigers to the local swimming pool. A German lady animal trainer has even managed to get on a similar footing with polar bears.

The great animal trainers of the last hundred years have

contributed more to the understanding of the more intelligent species of large beasts of prey than all previous books put together. Other factors which have helped to spread knowledge about the behaviour of lions, tigers, and other big cats, as well as bears, are the cinema and television—media available to the general public on an unprecedented scale. The reason for their immense popularity is of course that tame and 'well-behaved' animals go down well with the viewing public.

Lions and tigers do not only jump through burning hoops—they balance on balls in the circus ring, and allow people to ride on their backs, put their heads in their jaws, or use them as pillows.

One of the highlights of big cat training is tigers riding on the backs of horses or elephants, a feat first accomplished many years ago. One Russian circus had four tigers each on a different horse at the same time. One German trainer had trained one of his lions so well that his daughter sat on its back while it jumped from stand to stand—a trick which would be impossible, incidentally, with the less heavy leopards and jaguars. But when such creatures allow themselves to be wrapped round their trainers' necks like fur collars, or carefully remove a lump of meat given as a 'reward' from the bare chest of their trainer lying flat in the sawdust ring, the height of success in the training of these sensitive creatures is achieved.

Considered impossible about a decade ago, polar bears now jump through burning hoops. A 'mixed bag' of lions and tigers is led into the ring by a teenager. A tiger leisurely climbs up a flight of steps and then plunges headlong into a pool in which other tigers and their trainers are bathing together, after which they all leave the pool side by side.

How much further can the limits of adaptability be stretched? Big cats are clearly not only adaptable, but are always giving fresh evidence of their readiness—based on habit and confidence—to do things which in the wild they would not be required to do. They obviously feel quite happy in their friendly subordination to man.

Before people realized how docile and trainable these animals really are they depended for their knowledge on descriptions by explorers and big game hunters, through some at least of whom the big cats came to be regarded as bloodthirsty beasts with enormous claws and formidable canine teeth. In the long succession of hunting memoirs some accounts are outstanding because they throw light on the intelligent behaviour of lions and tigers.

CHAPTER THREE

WILY HUNTERS

WHEN stalking or lying in wait for their prey—antelopes, gazelles, zebra, etc.—lions and lionesses prefer to conceal themselves in tall, yellowish grass and undergrowth, which are ideal camouflage for them. They know the best hiding-places from experience of the runs where animals habitually pass, located near river banks and lakes and leading to the watering-places.

Big game hunters and wildlife observers jokingly assert that lions have so mastered the art of making themselves invisible that they could hide themselves out of sight behind a cigarette box on a billiard table!

Several lion families band together into a 'hunting party with division of labour' and set off into the bush to capture their fleet-footed quarry by tactical planning, which reduces their effort to a minimum and provides the smallest possible margin of error. C. G. Schillings, the famous traveller who returned in 1904 from an African expedition lasting several years and spent his leave in Berlin writing up his extensive notes, makes a number of very interesting observations in his book *With Flashlight and Gun*, published in 1905. *In the Wild and Captive* (1913) describes how some of the lions act as scouts while others form a circle of 'beaters'.

Their intelligent behaviour is discussed by the soldier and ornithologist Colonel R. Meinertzhagen, C.B.E., D.S.O., in his

Kenya Diary 1902–1906, first published by Oliver and Boyd in London in 1957, in which there appears the following account of one of his experiences in the bush:

> 20.*vii*.1904. *Nairobi*. The highlight of the day was watching a pride of lion hunting, a scene I had always hoped to see. It was about 4.30 p.m. in thin bush, over which we were able to have a good view from a slight eminence. The whole procedure was most deliberate. When we first saw the pride they comprised 2 lion, 4 lionesses and 3 half-grown cubs, and they were all more or less in a bunch and looking in all directions. About 500 yards from them and upwind was a herd of 15 zebra. From the stealthy movement of all the lion it was clear that they were on the hunt and that they had spotted the zebra; the two lion with two lionesses and the three cubs then made a wide detour, using every possible fold in the ground and bushes to keep hidden from the zebra, with the clear intention of getting round them and stampeding them by giving them their wind. The two lionesses left behind separated and took up crouching positions some 100 yards apart, both intently watching the zebra. Meanwhile the main body of lion had reached a position where the zebra should have got their wind. Suddenly up went the zebras' heads in alarm and they stampeded downwind, while the lions with cubs lay flat with only their heads erect. The herd of zebra passed between the two crouching lionesses but only some 20 yards from one of them; she lay very flat in a ready-to-spring attitude; it was most exciting. As the herd of zebra got more or less level with the lionesses they suddenly stopped and looked back. . . .

This is just how the males had planned the operation, and so the hunting party earned its reward. One of the lionesses sprang at the throat of the nearest zebra and brought it down. As the rest of the herd hesitated, panic-stricken, a second lioness, exploiting their confusion, seized another victim with lightning speed. Two zebras being enough for their needs, the lions allowed the rest of the herd to escape—an admirable and characteristic gesture of lions, which only kill to satisfy their hunger, thus ensuring a continuing supply of meat.

Lions also apparently learn from leopards, when their methods seem to them worth imitating. Leopards by no means restrict themselves to trees for lying in wait, but also stalk their prey in sparsely covered terrain. When, however, there are fair-sized trees near a run frequented by animals on which they prey they prefer to hide in the branches and pounce on their victims from a height.

During a hunting expedition through Tanganyika in 1926 the British explorer Arthur Loveridge saw a lioness spring with perfect timing from a thick bough to bring down a water-buck. In this territory lions and leopards live in fairly close proximity to one another. Loveridge was of the opinion that the lioness must often have watched leopards lurking in the trees and come to the conclusion that it was a method of hunting worth copying.

The lions of the African bush and the jungle (in so far as they venture into it at all) are much too clever to pick on elephants as adversaries. They keep well clear of them. In India, Burma, and Java tigers do not normally attack elephants either, unless they are carrying humans and the tigers feel in danger. Young elephants, on the other hand, are welcome prey to tigers if they are lucky enough to find one straying from the herd.

Similarly, African explorers have frequently observed lions giving rhinoceroses a wide berth. Stories of supposed attacks by prowling lions on rhinoceroses have never been supported by satisfactory evidence.

Lions will only very occasionally challenge even Cape buffalo. Their chief source of food is antelope and zebra, though they will also attack gnus, the largest of the antelopes in south and east Africa, and giraffe. Occasionally they will take domestic cattle, sheep, goats, and pigs.

It is clear, therefore, that they draw a sharp distinction between what constitutes a big risk and what they can kill with practically no risk to themselves.

The lion's courage is variously assessed by modern hunters. Impudence and cunning, with a dash of audacity, are found in older lions which, from laziness or convenience, turn into man-eaters.

A lion will not normally attack a man without a reason. So long as it can obtain the tender flesh of a zebra or an antelope it will not seek human flesh, the taste of which is unknown to it. But its attitude may change if on one of its forays it chances upon a careless black or white man and attacks because it feels

Photo: Ullstein

This English setter gives the impression that she is really playing the piano

Photo: laenderpress

For a fresh fish Flippy the dolphin will perform every trick expected of him

in danger. Then it will discover that a man can be more easily taken than any of the large horned beasts, which quite apart from their strength are able to outrun the lion.

Just how cunning lions—especially old lions—can be once they realize that human flesh tastes quite as good as the flesh of zebra is shown by the fact that they will remain in the vicinity of a native village where the chances of obtaining their meals with little effort seem good. Such a 'switch' is particularly common among lions which have grown too old to hunt down the fleet-footed antelope and zebra any longer. Similarly, lions which have sustained an injury that prevents them from hunting successfully but which retain the will to live in spite of the disability are liable to become man-eaters.

Such old and astute man-eaters were encountered by the celebrated African missionary and explorer David Livingstone in the years 1858–63. On one occasion he was amazed to find, in the middle of Africa, a village where all the huts were built on piles. These huts could only be reached by means of notched steps or rope ladders. The inhabitants had been forced to secure their dwellings against attacks by lions, which had decimated the village community. The surrounding region had an abundance of game, which the villagers hunted with their bows and arrows and spears. As soon as the villagers took their precautions the lions left the district, because they were powerless against continuous attacks from the raised platforms of the huts. For them, man had become as dangerous as the numerous Cape buffalo which grazed round about.

Livingstone supposed that these must be the same lions as had previously been seen unsuccessfully attacking a herd of buffalo south of the Libituane ford. There the enraged bulls, their heads lowered threateningly, had massed in a frontal assault on a pack of lions and put them to flight. One of the lions had been tossed and killed. This 'hunting accident' had taught the survivors to keep well clear of the buffalo. In their quest for other prey the hungry beasts had chanced upon two Africans and devoured them. After that they knew where to lie in wait for other human victims.

Lions generally stay close to villages because they are attracted by the cattle, sheep, and goats there. They realize that these copious stocks are a far better proposition that the wild animals roaming free on the vast expanses of open country. Sometimes they unexpectedly meet a dog, to whose flesh they are not averse, though they disdain the flesh of the hyena and the jackal, which belong to the same family.

In villages where lions had taken a heavy toll of human lives and cattle, and where they were not even deterred by the thorn hedges erected as a barrier against them, it was natural that the arrival of white hunters should be the signal for a great celebration. The villagers showered them with hospitality, in the expectation of being delivered from the menace of the lions in return.

And so it was that Livingstone was implored by the black headman Mebálwe to come to the aid of his threatened village of Mabota by undertaking a full-scale lion hunt with firearms. Livingstone agreed to do so—but during the hunt he was pounced on by a lion, as he describes in his book *The Zambesi and its Tributaries*:

> Being about thirty yards off, I took a good aim at his body through the bush, and fired both barrels into it. . . . I saw the lion's tail erected in anger behind the bush and . . . the lion just in the act of springing upon me. I was upon a little height. He caught my shoulder as he sprang and we both came to the ground below together. Growling horribly, close to my ear, he shook me as a terrier dog does a rat. The shock produced a stupor, similar to that which seems to be felt by a mouse after the first shake of a cat. It caused a sort of dreaminess, in which there was no sense of pain or feeling of terror, though quite conscious of all that was happening. . . . Turning round to relieve myself of the weight, as he had one paw on the back of my head, I saw his eyes directed to Mebálwe, who was trying to shoot him at a distance of ten or fifteen yards. His gun, a flint one, missed fire in both barrels. The lion immediately left me and attacking Mebálwe, bit his thigh. Another man, whose life I had saved before, after he had been tossed by a buffalo, attempted to spear the lion while he was biting Mebálwe. He left Mebálwe and caught this man by the shoulder; but at that moment the bullets he had received took effect, and he fell down dead.

In 1848 the African explorer Gordon Cumming set up base in

South Africa in a Bakalahari village. The following incident is described in his book *Five Years of a Hunter's Life in the Interior of South Africa.*

After communal supper one day, three of his men returned to their own fireside and lay down. Two of them, Hendrik and Ruyter, lay on one side of the fire under one blanket, the third, John Stofolus, lay on the other.

> Suddenly the appalling and murderous voice of an angry bloodthirsty lion, within a few yards of us, burst upon my ear, followed by the shrieking of the Hottentots. . . . Next instant John Stofolus rushed into the midst of us almost speechless with fear and terror, and eyes bursting from their sockets, and shrieked out, 'the lion! the lion! He has got Hendrik, he dragged him away from the fire beside me'. . . . It appeared that the lion had watched Hendrik to his fireside, and he had scarcely lain down, when the brute sprang upon him and grappled him with his fearful claws and kept biting him on the breast and shoulder, all the while feeling for his neck; having got hold of which, he at once dragged him away backwards round the bush into the dense shade. . . . The next morning, just as the day began to dawn we heard the lion dragging something up the river side under cover of the bank. . . . In the hollow where the lion had lain, consuming his prey, we found one leg of the unfortunate Hendrik, bitten off below the knee, the shoe still on the foot, the grass and bushes were all stained with his blood, and fragments of his pea-coat lay around.

Not far away the lion was found asleep in some undergrowth. Harried by Cumming's dogs, it made to flee but Cumming shot it dead. It was later identified as the man-eater which had long troubled the village. When it was finally confronted by a hunter with fire-arms, it was too late for it, used as it was to the primitive weapons of the negroes, to learn new tricks and readjust itself to different circumstances.

In 1900, during the construction of the railway which runs from the port of Mombasa on the east coast of Africa diagonally through Kenya north of Lake Victoria to Uganda, one of the many problems was the presence of lions which lurked in the neighbourhood of the workmen's huts and took the opportunity of pouncing on unsuspecting victims. Where the eastern border of the Amboseli National Park now runs, one of these lions would suddenly appear in broad daylight, and make off with the meat being prepared for roasting, and sometimes with whole carcases.

The following incident occurred near the small station of Kimaa, between the stations Machakos and Kibwezi, about 250 miles from Mombasa. One day one of the negro kitchen-hands fell a victim to a hungry lion concealed in the long grass. The man was dragged several hundred yards into the bush before being eaten. From then on the lion developed a great liking for human flesh, and within a short time thirteen other workmen in the Kimaa district were killed while out on their own. As a result, the natives refused to continue work on the railway, the Kenya Government offered a big reward, and the chief of the English railway police, a man named Ryall, was sent out to the trouble spot.

As work on the railway had been suspended for a whole week by the time Ryall arrived, the lion had taken to reconnoitring the near-by railway station for its next human victim. Ryall was able to observe it through his telescope, but could get nowhere near it, for as soon as anyone came into view carrying a gun the wily beast immediately withdrew, having learned from experience the nature of fire-arms.

One afternoon the hungry lion tried to force its way into the railway station, to attack the clerk on duty there. As all the windows had been barricaded the lion began tearing away the sheets of corrugated iron on the roof. When the terrified clerk telephoned for help several railway police came hurrying along, but just as they arrived the lion prudently beat a retreat, withdrawing into the bush to await its next victim.

A locomotive driver had concealed himself beneath the lid of a water tank with the intention of shooting the lion at the first opportunity through a slit in the tank. But the lion must have seen him take up position there; it crept up from the other side, sprang onto the tank, and tried to 'fish' him out with its paw through a hole in the top. A wild shot from the trapped man just missed the lion, but this was enough to persuade it that discretion was the best course.

The railway police thought it likely that the lion would make a further attempt a night or two later. Ryall, Sergeant Parenti and a German technician named Hübner took up positions in the police chief's coach, which was in a siding. In the next compartment to theirs three negroes had been posted in case their help was needed, one acting as look-out man while the other two slept. The first night one of them spotted a shadowy form prowling about, and shot at it. The next day and the following night there was no sign at all of the lion, and it was therefore assumed that it had gone off into the bush. On the third night it so happened that the police chief, whose turn it was to supervise the sentries, fell asleep on duty from sheer exhaustion. Outside, the ravenous, prowling lion must have heard him snoring on the other side of the sliding door, which had been left open because of the heat.

Terrible shrieks awoke the German on one of the upper bunks. Beneath him he could hear the crunch of bones and the suppressed growling of the lion. In his terror he jumped down into the pitch-dark compartment, landing right on the lion's back. The lion snarled but did not release its victim. Hübner, groping for the door leading to the next compartment, discovered that the terrified negroes had closed it and were leaning their full weight against it, in the belief that it was the lion that was trying to open it. Escape was therefore impossible for Hübner, nor could he locate his gun. He shouted to the negroes to stand back and let him through, and hammered on the door with his fists.

The lion, suddenly realizing that it too was trapped, leapt right through the window with Ryall in its jaws.

The other English policeman, who had escaped injury, lay trembling with fright on his bunk, not daring to move. In the dark, moonless night the lion was pursued by the light of hastily lit lanterns, but the search proved fruitless. Not until the following morning were the police chief's remains discovered, several hundred feet from the siding in dense undergrowth.

The man-eater lay low whenever a hunting party was sent out, but still remained in the locality. The negroes resolutely refused to return to work until the menace, now the terror of the whole village, had been destroyed.

At the far end of the station a stout box trap was installed and next to it a butcher's chopping-block was placed with a large lump of beef on it. As soon as it was all quiet and there was no one about, the lion came and took it. Next a juicy sirloin was placed in the trap, and again everyone withdrew. Soon afterwards a series of roars indicated that the lion had been caught in the trap. The man-eater was shot dead at point-blank range; and the negroes returned to work.

Before the Uganda stretch was completed in 1931, lions had killed a good many more workers. It was known as the 'lunatic line', though it was in fact a great pioneering achievement.

There are still man-eating lions in this region. The files of the Kenya-Uganda Railways contain telegrams demanding better security measures for railway staff threatened in the remoter stations by the continued presence of lions. A few years ago along this stretch two lions suddenly appeared in the path of a pointsman, who was obliged to shin up a telegraph pole. A station-master who had seen the incident telegraphed the next two stations up the line to the north, knowing that a southbound train was due to pass through them shortly. A squad of railway police, armed with carbines, climbed aboard the locomotive and saw the two lions leave the telegraph pole just as the train approached. They jumped down from the train, but were too late to fire at the lions, which disappeared into the tall grass. The pointsman climbed down, exhausted but relieved. A few days later the police tracked down and shot the marauding lions.

Tigers are equally astute tacticians, and work as a team, just as lions do. Von Rosenberg, a well-known explorer of Indonesia, collected a number of authenticated accounts of the guile used by tigers to acquire their human prey. They provide spectacular proof of these animals' understanding and co-operation among themselves.

In 1854, in a district of Malang on the island of Java, not far from the villages of Songoro and Gondangleqi, in a region infested by tigers but inhabited by very few natives, five Javanese returning from a fishing expedition decided to spend the night in an old deserted *gobok*—a hut built on stakes serving as a look-out post to protect crops against predatory birds and animals. The actual hut rested on four strong bamboo poles. The platform, some twelve feet from the ground, was reached by means of a retractable ladder with bamboo rungs.

As the full moon rose over the valley the fishermen heard a sudden roar just a few feet away, and were terrified to see a tiger standing on its hind paws clawing at the hut. But knowing that tigers are incapable of climbing up smooth poles, they were reassured. Apparently aware of the fact itself, the tiger stalked off.

The exhausted fishermen settled down to sleep on the floor of the hut, except for one who kept watch. After about half an hour the sentry, frightened by the renewed roaring of a tiger, raised the alarm. In the moonlight the men saw, to their horror, *two* tigers three hundred feet away approaching the hut. On reaching it they went across to opposite poles and reared up on their hind legs, hissing. Striking their huge paws against the poles, they tried to bring down the hut. Four of the terrified Javanese were rooted to the spot with terror, but the fifth, the leader of the party, jabbed at the tigers with a piece of wood, whereupon they made off in the direction of the river.

After a while both tigers reappeared, this time accompanied by a *third*. They made their way straight over to the hut and each went to a pole and tried to shake the hut down. The attempts of the desperate men to ward off the tigers were successful. One of the tigers was wounded in the left eye by a stake which one of the men had sharpened during their short-lived respite, and quickly retreated into the darkness. In trying to stab at the others, however, the leader had dropped the stake, and was reduced to using his *gollok*—a short dagger-like sword—and a long bamboo stick, the only weapons he had left. One of the men had the idea of throwing the baskets laden with fish down to the tigers, in the hope that they would appease their hunger with them and then withdraw. The tigers smashed the baskets to pieces but disdained their contents, renewing their fierce onslaught on the hut. The continual shaking of the hut caused the bamboo ladder to go hurtling down; it hit one of the tigers, which immediately ran off. The remaining tiger seemed to realize that it was impossible to continue unaided, and went running after its companion.

The men optimistically believed that the tigers would now give up and leave them in peace. Fleeing was out of the question, as they were too far from Songoro, the next village. They therefore had no option but to stay put till morning.

Suddenly Sarno, the youngest of the party, gasped "Matjan! Matjan!" (Tigers! Tigers!). Now *four* tigers were clearly visible in the undergrowth, bounding over to finish the destruction of the hut. Each tiger made its way over to a corner pole, and a few minutes later the hut collapsed.

Four of the men were hurled sideways, and each of these was seized by a tiger. Three of the tigers immediately dragged their victims off. As the hut collapsed, the fifth man ended up lying beneath the roof and so escaped the remaining tiger's notice. When he recovered consciousness he felt his *gollok* beside him, and the bamboo stake which had fallen to the ground also within hand's reach.

A few yards away the tiger was tearing the fourth young man, already dead, to pieces. The survivor stabbed at it with his stake, but it merely pushed it to one side and snapped at it. The fisherman persisted and with a desperate lunge stabbed the creature through the heart with his *gollok*. The tiger keeled over, dead.

The fisherman, utterly exposed as he was to the attacks of other tigers, overcame his fear of the dark jungle, and climbed up the first tree he could to await the dawn thirty feet above the ground. After walking for several hours, with his *gollok* in one hand and the bamboo stake in the other, he finally managed to reach Songoro, unmolested. There he met a number of Boers who had arrived there a few days previously on their travels through Java and Sumatra. Accompanied by villagers, they made their way to the scene of the slaughter, where they discovered the dead tiger alongside the fisherman it had torn to pieces. Following the tracks of the other three tigers, they soon found the remains of the other men. The tigers themselves were nowhere to be seen.

Von Rosenberg heard this account from the Boers. On his return to Germany he gave the details of this gruesome incident to the Zoological Society, which published them. Europeans who have been officials or merchants in Burma, India, and Indonesia, as well as explorers and hunters, have provided a great deal of evidence of such 'planned operations' by tigers, which explains why even today tigers are still greatly feared.

CHAPTER FOUR

PET CATS, BIG AND LITTLE

IT is not only at circuses that tamed wild animals display their ability to perform tricks. They also show themselves to be highly adaptable and intelligent if brought up from a very early age in the constant company of human beings.

Young lions in daily contact with people who provide their food and give them a carefree life grow up to be much more amenable than most leopards or tigers reared in the same way. The lion's perceptiveness soon helps it to understand that a friendly relationship with its provider can only be to its advantage. This is often just as true of lions captured in the wild as those reared by humans from an early age. With intelligent handling they become as tame as some of the ancient Pharaohs' lions are said to have been, or those belonging to the Carthaginian military leader Hanno or to the Roman Emperor Elagabalus.

Brehm made some interesting discoveries with Bachida, his tame lioness, by rearing her at home, and similar results have been achieved by many other owners.

During his two-year stay in Nubia, Bachida proved to be an excellent example of the extent to which such intelligent creatures can adapt their mode of life to frequent contact with human beings. "Bachida accustomed herself to her new surroundings", wrote Brehm. "Soon she was following me around like a dog, showing signs of affection at every opportunity and proving quite a problem from time to time by coming into my bedroom and waking me up with her caresses in the middle of the night. After a few weeks she had become the undisputed mistress of all the animals on the farm, but was much more interested in playing with them than doing them any harm."

Only one creature on the farm claimed Bachida's respect, and that was a completely tame marabou stork, which on one occasion had drawn blood by pecking her with its wedge bill. Ever after she had given it a wide berth.

Bachida accompanied Brehm on his expedition to the Nile and went with him to Cairo. There he took her through the streets on a lead, just as Queen Berenice had taken a pet lion around the city over two thousand years earlier. During the voyage from Alexandria to Trieste Bachida walked about on deck with her master, which gave the passengers an unusual topic for conversation. Brehm then handed Bachida over to the Berlin Zoo. When he went to see her there a couple of years later she immediately recognized him.

Some fifty years later, the Duchess of Montgelas acquired a cub lioness named Cleo, and gave her a frisky fox-terrier as a playmate. She became so attached to it that she sulked whenever she was separated from it. For a time Cleo lived in a courtyard on the Duchess's estate with several dogs, a tame roebuck, and a baboon. All these creatures quickly realized that Cleo was quite easy to live with and constituted no danger to their well-being. The baboon was fond of going for rides on her back, taking the opportunity while he did so to inspect her fur, as monkeys do among themselves, for the grains of salt which they enjoy eating.

When Cleo was allowed to share her mistress's company indoors she would stretch herself out full-length on the sofa, the softest and most comfortable place in the house. She treated the sofa cushions as excellent playthings and was rather rough with them; when the Duchess wanted to remove one it usually led to a tussle, but a sharp word or a box on the ears was enough to make it clear who was mistress of the house.

Cleo had one peculiarity which disturbed her otherwise very peaceful and pleasurable existence—she could not stand her mistress playing the piano. The moment she heard the first notes she would come running up and poke the Duchess in the ribs, then put her forepaws on her lap, and, if that did not make her stop, she would carefully take one of her mistress's hands in her mouth and gently pull her away from the piano. How the Duchess ever managed to practise is a mystery. It is well known that dogs are also liable to sulk or rebel when their masters play the piano.

This unusually talented lioness was full of curiosity. In the courtyard and garden her behaviour resembled that of poodles, wire-haired terriers, and most other dogs, including hunting-dogs like dachshunds. Familiar farmyard noises—the cackling of hens, the crowing of the cock—and the footsteps of the labourers did not bother her in the least. But whenever she heard the footsteps of the postman, or if a visitor arrived with a strange dog, or if a cart went trundling by, Cleo would dash over to a window to see who was coming, though she never jumped out, whoever it was.

If her friend the fox-terrier was in the garden Cleo would get very excited. If he came into the room she would play with him, allowing herself to be nipped in the tail or the ears; then, lying on her back, with her claws sheathed, she would raise the terrier high into the air, causing it to yelp with delight. If the dog thoughtlessly nipped her too hard she would growl a warning, but instead of retaliating, as a cat probably would in similar circumstances, she would go running to her mistress to 'complain' about the dog's rough play. She seemed to know that she must not give tit for tat, as this, with her vastly superior strength, would have annihilated her little playmate.

Young lions, like those which Helen Martini reared at home during the period she was a keeper at the Bronx Zoo in New York, are just as adaptable as dogs. She always had them around her, even while doing her household chores. In her book *My Zoo Family* she throws interesting light on the playfulness of her lion cubs and their willing submission to her in her role as a sort of foster-mother. They developed into lions capable of being taught a number of tricks, a character they retained on being moved to the lion house at the Bronx Zoo.

Joy Adamson, in *Born Free*, gives a perceptive and penetrating account of Elsa, the lioness which she reared. The cubs at the Bronx Zoo were in the habit of congregating at the door of the cage while awaiting visitors, and Elsa opened doors herself in order to lose no time joining the company of people she knew. She would obey a variety of verbal orders and reacted like an obedient dog to a sharply spoken "No!". Later she was given her liberty, but she would always return of her own accord to the friendly people she knew.

In a second book Joy Adamson tells how one day, after a long absence, Elsa suddenly appeared with her cubs among her human friends, entrusting them to their care with complete confidence. And, like Elsa herself, these cubs developed into intelligent creatures with surprisingly acute perceptive faculties. They got to know what they could and what they could not do if they meant to make the most of the advantages offered them by living among human beings.

Experiences like those described by Helen Martini and Joy Adamson bear out what Brehm experienced with his lioness Bachida and the Duchess of Montgelas with Cleo. They are confirmed by zoo directors. The behaviour of lions therefore, in spite of the fact that they belong to the 'big cat' tribe, is comparable, even when fully grown, to that of faithful dogs, provided they are reared among humans who treat them with kindness.

Sometimes when the larger members of the cat family have young they reject them, and in such cases the cubs are reared with a bottle or given into the care of a canine bitch.

Dr Knottnerus-Meyer, former director of Rome Zoo, has recorded his observations of such cubs reared at his home. The first two which he adopted later became so utterly devoted to their keeper, after they had been transferred into cages, that, at a given word of command, they would surrender their meat to her—something they were not prepared to do even for the director, with whom they had spent the first five months of their lives. Dr Knottnerus-Meyer modestly explains: "They allowed their keeper to remove the meat from their jaws. I myself could not venture to do this, as not being the one who fed them I did not count for so much to them."

Twice a day the two cubs were carried from their play-pen

into the house in the arms of the director or the keeper to receive their ration of meat and milk. Two further meals were brought to them in the play-pen, but they pined to be indoors, where they could play in the immediate presence of their human friends and where, because they looked on their foster-mother as their real mother, they felt infinitely more at home. They therefore looked for a way of getting their other two meals in the house too—and after giving a good deal of thought to the matter they succeeded. The improvised play-pen was surrounded by a wire-mesh fence four feet high, which in the view of those who had installed it was quite high enough to prevent their breaking out. But it was not long before one of the cubs discovered that to climb up the wire he had only to support himself against the trunk of a maple-tree, growing just a foot or so from the fence. And so he managed to reach the top of the wire, at which point he simply let himself topple over on to the grass on the other side. His equally intelligent sister was not slow following him. Then together they made their way straight over to the house. Dr Knottnerus-Meyer comments: "There they were, standing one day outside my door waiting for their breakfast. They had been waiting quietly and patiently till someone came to let them in."

After this the two cubs often went off on their own for a tour of the zoo, mixing freely with visitors. For their favourite spot they chose a mound behind the zoo wall overlooking the hustle and bustle of the street beyond. When they grew older they naturally had to be confined in escape-proof cages. It is extraordinary that, even at the age of three, they both still gave a rousing welcome to their keeper, even though they saw him every day, and they still grew very excited whenever they saw their 'foster-mother', and would not calm down till she had entered the cage and played with them for a while. It says much for their understanding of things that they never once caused her any injury and deliberately kept their claws sheathed.

Dr Knottnerus-Meyer also owned a tame African cheetah. This got on so well with one of the lions that in time it became possible to present them both to Princess Radziwill in Rome at at the ballroom of the Hotel Excelsior during an Easter carnival. The cheetah took its place next to the Princess in a small Roman chariot drawn by a sturdy pony, and behind walked the keeper with the lion, which eyed the seething mass of revellers with some misgiving and seemed dazzled by the lights of the ballroom chandeliers. But the cheetah was obviously enjoying every moment, for it was used to being paraded in an open carriage though the streets of Rome, without ever causing the least trouble.

A lion suckled and reared by a dog usually becomes so attached to its foster-mother that if it is taken away from her the separation has noticeably adverse effects on its subsequent development. E. Wells, in his book *Mit Löwen auf Du* (1935), tells how a mongrel foster-mother which was expecting a litter, having been temporarily removed from the lions' cage, had to be put back there almost immediately because they became terribly agitated at night without her. After the bitch had given birth to her pups and had looked after them for a few days, leaving them for only short periods during that time, she spent the following weeks sharing her time and attention between her own offspring and her adopted lions, but at night she always stayed with the lions, who were then no longer restless.

In a number of circuses, as well as Leipzig Zoo, dogs have actually been left in lions' cages to have their litters, and Willy Hagenbeck went so far as to install a bitch about to give birth in a cage occupied by tigers to which it had been foster-mother. Everything went smoothly; none of the tigers attacked the pups, but on the contrary were visibly pleased about the 'additions to the family'. Something well known to antiquity but regarded in more modern times as impossible was thus confirmed. Visitors to the zoo were able to enjoy the delightful spectacle of these huge beasts of prey playing gently with the pups and, without a murmur of disapproval, tolerating the cuffs and bites inflicted on them.

Many cat-lovers have tried to train their pets to perform little tricks like those seen at circuses performed by lions and tigers. With enough patience and systematic training the results can be quite successful, domestic cats being in no way inferior to their larger relatives.

In 1932 I owned a tabby cat which would come up to me of its own accord to have its harness and lead fitted whenever it saw that I was about to leave the house. Its harness consisted of a red leather jacket with straps, similar to those worn by dogs. The lead was not fastened directly to the collar, as is usual, but to a ring on the jacket half-way along the spine. One day I had the harness and lead all ready on a chair, but was delayed for a while waiting for the postman. The cat showed its impatience by jumping up at me and then taking the harness in its mouth to make it quite clear to me that it was high time we were on our way. Sometimes it would go and sit under the harness, which hung on a peg in the hall, and look up at it wistfully as though to say what a good idea it would be to go for a walk, or better still a ride on the tram.

It would offer its right paw to visitors when they called or were about to leave. In time it extended its repertoire by jumping over sticks held out at arm's length and springing through hoops. At first it would only perform its hoop trick from a position off the ground—for example, from a chair or table. It came to enjoy these games so much that it would come running over whenever called and perform them without any prompting from me. It later learned to leap through two hoops, but these had to be not more than about eight inches apart.

The idea of experimenting with such tricks with my cat was suggested to me by an article in a newspaper about training dogs, in which the writer maintained that cats, unlike dogs would never of their own free will apply themselves to such antics. Some time before, I had seen at a circus touring Austria a group of trained domestic cats which performed exactly the same kinds of tricks as lions and tigers do. I therefore put the matter to the test, with the successful results described.

That domestic cats can be trained has been known for a long time. As early as 1753 performing cats were being exhibited in Paris, and since then menageries have presented shows in which cats and dogs and other animals—even rabbits, hens, and rats—have performed together.

In 1882 a fatal accident occurred during a performance of trained cats in the French village of Beaupré-sur-Saône. The owner had painted his fully grown cats to resemble young tigers.

They were brought into the ring by a stunted country lad of sixteen who had been dressed up like a lion-tamer. At one of the performances the large audience saw the biggest of the cat troupe, which evidently felt that it was being maltreated by this undersized youngster, crouch menacingly, then spring at him and bowl him over. All the other cats joined in and attacked him so savagely that he died. It was the sort of surprise attack which lions and tigers will occasionally make.

One is always hearing stories about cats which develop special aptitudes. In 1955 there was a cat which played bar billiards with the customers of a public house in a German village, using its paw as a cue, and winning lots of prizes in the form of meat-balls.

In another part of Germany a tom-cat earned fame by sharing the spoils pilfered from the local butcher's shop or its own master's larder with its best friend, a German sheepdog, in return for which the dog protected it from being chased by other dogs, besides manipulating door handles with its paws for its friend to break in and steal meat.

Thirty years ago I owned a cat which adored riding in my car when we went out visiting. In other people's houses it would sit patiently for hours till it was time to leave. This cat had taught itself to open doors by jumping up and pressing down on the handles with its paws. Inquisitive by nature, as all cats are, it liked sniffing around in all the rooms of the house. Even if it lay curled up in its basket when I returned home, the fact that a number of doors were open showed that it had been on a tour of inspection, for cats find it impossible to close doors after them.

The Duchess of Montgelas, whom we mentioned earlier, had a cat which grew up with a pig-tailed macaque monkey. It enjoyed playing with the monkey, and even allowed it to share its food, but if it tried, as monkeys do, to take food out of the cat's mouth, the cat would get very angry and hit it on both ears with its paws. The cat did not, however, object to the monkey then inspecting its paws; but knowing that the real interest lay in seeing how the claws fitted into them, it kept them sheathed, in any case not wishing to scratch its friend. The monkey, find-

Photo: Associated Press

Betty Goodson, a young teacher, and schoolchildren in New Zealand playing with 'their' dolphin Opo

160

The tame pelicans and dolphins of Marineland in Florida are on the best of terms *Photo: Ullstein-U.P.*

Flippy loves playing basket-ball *Photo: Ullstein-Popper*

ing no claws, then bit the cat, which retaliated by again boxing its ears, so that it was finally persuaded that it had better behave itself. But the cat never bore it any grudge. It would climb into the topmost branches of the trees with it, and keep it company when it was chained up.

When the monkey died this cat transferred its affections to a wire-haired terrier, with which it shared its basket even while having kittens, as a result of which they came to regard the dog as their father. Convinced of the 'foster-father's' kindness towards her young, the cat would leave it to look after them while she went off for a stroll in the garden. She even taught the dog the right way to arrange kittens in a basket to keep them warm. As they grew up, however, the kittens became rather too spiteful in their games with the dog, biting it with their sharp little teeth, until it could take no more and kept away from them. But this did not affect its friendship with their mother in the least.

PART FOUR

'TALKING' DOGS AND 'THINKING' HORSES

CHAPTER ONE

'BOW-WOW' WAS NOT THE ONLY WORD

THE earliest reports of a 'talking' dog and of other animals, such as cattle and sea-lions, having the power of speech date back to Pliny, among whose many and varied interests science held pride of place. He is highly sceptical in his reference to the talking dog in Book VIII of his *Natural History*: "A dog is said once to have spoken, but be it noted that such a thing would be a miracle." A few pages further on he reports: "On a number of occasions our forefathers experienced the miracle of hearing oxen speak. As soon as this became generally known, the Senate was obliged to convene in the open air." A later Roman historian, Valerius Maximus, discussing the Second Punic War, says: "When Publius Volumnius and Servius Sulpicius were consuls, a remarkable phenomenon caused great astonishment: an ox, instead of lowing, spoke like a human being." Another source tells of an ox that, during the same war, warned Gnaeus Domitius in a human voice: "Woe, woe unto you, Rome!"

To these altogether legendary reports from the third century B.C. another from Pliny may be added to show how current the idea was in ancient times that some animals have the gift of speech: "If sea-lions are addressed by their names they reply in a loud voice."

In a popular nineteenth-century anthology of writings on animal observation, *Museum des Wundervollen*, there is a story of a sea-lion put on show at the fairs which was able to pronounce the words "Yes", "No", "Mama", and "Papa" in the presence of its trainer. Such a report would seem to confirm Pliny's statements about the 'answering' sea-lions. It is worth noting in this connection that Carl Hagenbeck, founder of the famous animal park, wrote in the *Hamburger Fremdenblatt* on November, 15th, 1910: "Since I have had a walrus which can say 'Papa' loud and distinctly I am no longer surprised by anything."

In the Middle Ages, and even the beginning of modern times, which date back to Columbus's discovery of America, it was dangerous in some countries to own a dog which had, or was claimed to have, the power of speech. Learned people, but most of all superstitious townsfolk, peasants, and artisans, held the view till well into the seventeenth century that in such matters the powers of good and evil were at work. Many a respectable woman who gossiped about her 'talking' dog was burned as a witch.

Even in the time of Louis XIV people tried to account for strange phenomena in terms of the irrational and any poor dog credited with being able to speak was treated most unkindly. In the early seventeenth century there is an account of a whippet belonging to an Italian nobleman, Count Cirao, which could pronounce fifteen Italian and eight French words. It was killed by the peasants who declared it to be possessed of the devil.

At the beginning of the eighteenth century several reports of talking dogs came to the public's attention, but it was not until the philosopher Baron von Leibniz, whose name and status commanded general esteem, spoke out on their behalf that people began to adopt a more tolerant attitude towards them.

To begin with, about the year 1702 it was publicly announced that an Austrian had arrived in Amsterdam and Utrecht "who had with him a talking dog which could repeat nearly all the letters of the alphabet, with the exception of L, M, and N". Although we do not know which breed of dog this was, we do know that a dog belonging to a town councillor a century later which was said to be able to talk was a pug, a breed favoured by the Royal House of Orange, which William III (William of Orange) introduced to court circles in London during his reign. The town councillor in question, a certain Friedrich Krumbholtz, who lived in a small hamlet in East Prussia, had obtained the pug from a Königsberg merchant with commercial contacts in the Netherlands. In all probability the pug's new owner was some kind of practical joker, for he claimed to have got his "grey pug, an English breed, to express many words by stroking its throat!"

About this time reports of talking dogs became more and more frequent. One such report even found its way into the records of the Paris Academy of Sciences in 1706, linked with the name of Leibniz, who for six years had been President of the Berlin Academy of Sciences, created by him on the French model. He had told the Abbé de Saint-Pierre of "a farm dog of ordinary build and medium size" which could speak no fewer than thirty words. This dog lived near Zeitz and had been taught by a young boy. Apart from German words it had also 'mastered' a number of French words only recently absorbed into the German language (*Assemblée, Café, Chocolat,* and *Thé*). When it began to speak the dog was three years old, and its voice sounded "like an echo". The writer of the report lays great stress on the fact that "without such a source as Herr Leibniz as an eyewitness" he would not have had the temerity to report on the talking dog, and he ends his description by emphasizing again: "Once more then: Herr Leibniz has seen and heard it." Leibniz also mentioned this dog in a letter to the scholar Grimarest.

As the news of this dog was given in the weekly journals and spread through all the German duchies and principalities, dog-owners from far and wide began to assert that they too owned dogs which could articulate words intelligibly. One such animal, from Regensburg, was said to be able to pronounce the words *Thee* (tea), *Coffee, Chocolade,* and clearest of all *Monsieur.* Judging by the vocabulary, this dog may have been the same as the farm dog from Zeitz; so perhaps was a dog exhibited in 1722 in Regensburg "which could speak German, French, and English, and could also recite the whole alphabet", though if this were the case the dog must have achieved the advanced age of eighteen.

About 1720 there was a report of a dog as gifted as the one described by Leibniz. According to J. C. Fritschius, who in 1730 published a book of *Strange but True Stories concerning*

Theology, Law, Medicine and Physics, this dog had been reared on a country estate belonging to Herr von Hörnig near the small town of Weissenfels in Prussian Saxony. A young farm-hand had assiduously taught it to imitate the human voice by holding its neck tightly between his fingers and then jabbing them into its neck—a manipulation aimed at making it easier for the dog to articulate words.

On February 23rd, 1721 this same dog was sent as a rarity from its farm to His Highness Ernst Augustus at Weimar. There is no doubt that the Grand Duke of Saxe-Weimar must have derived a great deal of amusement from this dog, for it was able "quite distinctly to answer the following questions: Question: How are you? Answer: *Wohl!* (Well!). By what name are you known? *Spitzbube!* (Rascal!). Who is your father? *Ein Hund!* (A dog!). Who is your mother? *Eine Beize!* (A mongrel bitch!). What do you eat? *Braten und Fleisch* (Roast meat and raw meat). What have you learned? *Stehlen* (To steal). Where ought you to be? *An den Galgen* (On the gallows)." If at first it should appear that the Weissenfels dog was the same one as Leibniz referred to, the considerable differences of vocabulary indicate that the two descriptions relate to different dogs.

Something similar was reported some two hundred years later (*Hamburger Fremdenblatt,* March 30th, 1911) in connection with the training of Don, the most famous of all talking dogs. His gamekeeper owner was censured for stimulating articulation "by massaging the dog's larynx". And in the *Münchener Neuesten Nachrichten* (November 10th, 1910) there was a report of a similar kind of stimulation given by Professor Graham Bell to his terrier, which was credited with learning several words of English. Well known as the inventor of the telephone (1872), Bell was also a physiologist and wrote a book on speech therapy, in which he maintained that many animals possess the ability to utter articulated sounds, a process which can be encouraged by suitable manipulation. This 'throat therapy' enabled his terrier to articulate correctly many words it had learned. Clearest of all it said: "How are you, grandmama?" So Fritschius was on the right lines in his description of the methods used by the farm-hand with the dog from Weissenfels, thus bringing within the realm of credibility the phenomenon of a talking dog.

Following the wave of reports about talking dogs in the first two decades of the eighteenth century in Germany, there appeared in January 1829 an article in the *Dumfries Journal* about a dog living in London which had the power of speech and was said to have distinctly imitated the name of a friend of its owner: William.

In the first decade of the twentieth century there was a total absence of reports on talking dogs, but in 1910 another flood of published material appeared on the subject, unleashed by the "four-legged phenomenon", a German short-haired pointer named Don.

CHAPTER TWO

DON AND OTHER FAMOUS 'TALKERS'

Don had made his début in Hamburg, at the largest hall in the city. Tickets were as scarce as gold dust, although the hall seated 12,000. The Hamburg newspapers had written column after column about the 'wonder dog', while photographers and sketch artists had vied with one another to fill the city's newspapers and magazines with illustrations to accompany the articles. And now, on a memorable afternoon in the year 1911, the Berlin Press was all agog in the reception hall of the Hotel Bellevue awaiting the performance of the brown German pointer in company with Fräulein Ebers, the owner's daughter.

Don was not at all a prize-winning representative of his breed, having a rather stocky look about him. His shoulder height was about two feet, the colour of his short hair, dark brown, with an off-white stripe on his chest.

Professor J. Vosseler, at that time director of the Hamburg Zoo, which since the establishment of the nearby Hagenbeck animal park had lost most of its significance, reminded his audience in an opening speech about the newspaper articles which had appeared in 1910 relating to dogs which had been successful in uttering two or three words. Then the gamekeeper's twenty-four-year-old daughter, a fashionably dressed brunette, turned to Don and said: "You will now tell the gentleman your name—what is your name?" Don inclined his head a little, and in a hoarse voice replied, with gaping jaws: "*Don!*" Fräulein

Ebers put another question: "When you look at the gentlemen here sitting with their coffee and cakes, perhaps you too feel thirst or hunger?" The reporters watched attentively as the dog now strained its larynx to greater exertions and said quite distinctly: "*Hunger!*" "Good, but what would you like to eat now —a piece of bread, some sausage-meat, or would you rather have some cake?" Don again raised his head and gurgled in a clearly audible and unmistakable combination of vowels and consonants: "*Kuchen!*" (Cake). As though determined to remove any shadow of doubt Fräulein Ebers then asked: "So you would rather have cake than bread?" Having given every sign of listening intently, Don, opening and closing his mouth to the rhythm of the words, reaffirmed: "*Don Kuchen haben!*" (Don have cake).

After he had said this it was clear even to the most sceptical that this was no sham jiggery-pokery but that the dog really could talk; even if the pronunciation was reminiscent of the hoarse voice of an intoxicated man, the fact remained that his utterances were indisputably articulated human words. To conclude this press conference unique in the history of animal behaviour, the journalists were invited to converse with Don themselves. In so doing, they confirmed that Don was anything but a dog of few words, and was only too willing to display his acquired vocabulary.

From Fräulein Ebers they learned that it was over a year since Don had surprised her father with his gift of speech. Whenever he had given Don, or his other dog, a dachshund, a bone or some other choice morsel, he had always done what thousands of other dog-owners do as a matter of course, namely ask the question: "Will you have this?" To his astonishment Don had answered quite distinctly on one occasion "*Haben!*" (Have). He thought at first that his ears had deceived him, but Don firmly repeated the same word.

A few weeks later, annoyed at the din his dachshund was causing with the hens in the yard, he had opened the window and called to the dog "*Ruhe!*" (Be quiet!). Seconds later he heard just behind him a sort of echo of the word. There was no-one else but him and his dog in the room, and as he looked round he was flabbergasted to see Don raising his head, opening his mouth, and repeating through the window "*Ruhe!*"

In the months which followed the family concentrated on teaching Don to develop his extraordinary talent for imitating. The result was surprising, for Don increased his vocabulary to eight words, which at times, in the way budgerigars sometimes do, he strung together to form complete sentences, though not always using the words in a meaningful sequence.

Soon the gamekeeper's close friends were invited to witness Don's rare gift, and as the news spread through the village and beyond, hundreds of people got to know of the talking dog—though at this stage he was not receiving nation-wide publicity. It was not until Fräulein Ebers' fiancé, who was an editor, came to visit the family that things began to happen. On his return he published an article in the *Graudenzer Zeitung*, which was reprinted in a large number of newspapers in Germany and in other countries. This was in 1910. As a result, reports streamed in from dog-owners everywhere claiming that their dogs could speak one or two words. Intrigued by all this publicity, a theatre manager became interested in Don's talents, and having satisfied himself of the truth of the reports, engaged Don for a series of public performances.

The reporters in the Hotel Bellevue wrote voluminous articles accompanied by pictures of Don, resulting in a fantastic rush for tickets to the Wintergarten, where Don was to perform. The Company which owned both the Hotel Central, in whose complex of buildings the Wintergarten was situated, and the Hotel Bellevue, where at first Fräulein Ebers was accommodated with her 'wonder dog', had on some days to put hundreds of entrance tickets aside for hotel guests, as people were now arriving in Berlin from far and wide to see the most amazing dog ever with their own eyes and to hear him speak. Ticket touts did a roaring trade.

The Wintergarten was generally very well patronized because it offered first-class attractions, but it was very rare for guests to leave in their hundreds immediately after an act, without waiting to see the rest of the show.

Both for the hotel and the Ebers family a period of prosperity now began. Fräulein Ebers was paid 12,000 marks for appearing at the Wintergarten for a month. Similar amounts were paid for her appearances in Vienna, Moscow, and other European cities.

In the U.S.A. she received over 100,000 dollars. She and Don arrived back in Germany just before the outbreak of the First World War. During the war years Herr Ebers was careful to ensure that Don did not forget any of his eight-word vocabulary —in fact he discovered that Don had added to it, during his Russian tour in 1912, the word "*Rubel*" (rouble). But Don's age was now clearly beginning to tell on him, for when he found the noise of other dogs around him intolerable, instead of saying "Be quiet!" he said "*Rubel*"—a sure sign that his mind was beginning to wander.

Meanwhile newspaper editors were receiving sacks of letters about talking dogs, following the publicity which Don had received. The London *Daily Mail* investigated a report that a certain Dr Molden of Plumstead had an airedale terrier which could talk. The investigation revealed that the dog in question, which answered to the name of Buller, could say a few set phrases, but came out with them regardless of whether they suited a given situation or not. Buller had been taught to greet people with the words "God Save the King!" and it was claimed that he could also say "God Save our Gracious King!" He was also said to raise his right paw in salute as he spoke these words. If he was annoyed, Buller would first growl and then threaten "I'll tell Mama!" He is supposed to have had special difficulty pronouncing the word 'gracious', to do which he would twist his neck round, in much the same way as mynah birds do when imitating difficult words. In 'King' the K was said to be scarcely audible. According to Dr Molden, Buller could also say certain other words in a more or less comprehensible way. The Berlin newspaper *Der Tag* was so impressed with the *Daily Mail* article that it reprinted it on December 4th, 1910. The *Tägliche Rundschau*, a rival newspaper, feeling it had been scooped, went in quest of a talking dog of its own. It came up with a pomeranian which, though evidently less gifted than the airedale from Plumstead, would, according to its German owner, clearly articulate the word "*Hunger*" when feeding time came round.

In 1911 talking dogs as a current topic may be compared with flying saucers in the 1950s. In quick succession came two further reports, from North German villages, both vouched for by local dignitaries.

The first, published on November 24th, 1910, in the *Königsberger Allgemeine Zeitung*, concerned a brown short-haired dachshund called Willutzki, which belonged to the Mayor of Niedersee, who took it along with him on the local hunts. Willutzki had been trained to drive badgers, foxes, and rabbits from their holes. But he had also been taught to speak, and had proved to be quite an apt pupil. First he learned to bark the word "*Bitte*" (Please), later to pronounce his name distinctly, holding his head at an angle as he did so. When asked his name and what he was up to, he would whimper, "*Willutzki—lügen!*" (Willutzki—telling lies). He had also been taught to mark such occasions as the Fire Brigade's annual procession or the Emperor's birthday by barking out: "*Hoch, hoch, hoch!*" (Three cheers!).

The other report concerned a poodle which had been trained by its owner to answer "*Hurra!*" to the question: "What should a good patriot shout?" It is not clear from the newspaper accounts whether this dog had any other words in its vocabulary.

Many other dog-owners claimed the power of speech on behalf of their pets. In 1910 a terrier was presented at a Montmartre exhibition in Paris. The trainer explained to the audience which words the dog would say next, and then in some way manipulated the dog's nose. With just a little imagination the audience were able to make out such words as '*Au revoir*' (Goodbye), '*Bonsoir*' (Good night), '*Maman*' (Mama) and '*Papa*'. The entrance fee for this little show was twenty centimes. At least one newspaper agreed that it was worth it.

In the U.S.A. in 1922 there were reports of two dogs which could say several words. One of them would let its wishes be known by saying to its master, "I want to go" or, more insistently, "Go walk!" The other gave its master hints by declaring, "I'm hungry!"

In 1936 there was in the Netherlands a female sheepdog which, if she wished to share her mistress's meal, would look up at her and say in deep guttural tones "*Mama—hebben*" (Mama—have). The Dutch researcher into animal behaviour Bierens de Haan confirmed that this dog would answer questions by saying "*Ja, Jeanna!*" And Dr Bernhard Grzimek commented that it was a question of imitating the vowel sounds, which the dog did better and better the more it practised, and that the pitch and volume also improved with practice.

A dog which could say a few words was televised a few years ago in the U.S.A. and in 1955 there was a talking poodle living in a town in the Ruhr, but the sounds it produced were so indistinct that they could hardly be called an imitation of human words.

The zoologist Peter Kuhlemann reported in 1962 on two dogs which he had personally put to the test. One was a black and the other a white toy poodle, belonging to a certain Werner Fürst who lived near Kiel. Both followed a routine at meal-times similar to that practised by the pointer Don. Andrea and Xoxi, as they were called, obeyed commands to sit in their places in an arm-chair, and when asked the name of their mistress both replied very clearly: "*Mama*". Any deception can be ruled out, as the tape-recorder reproduces the sounds clearly articulated. These poodles could also dance, sit up and beg, and bring slippers or the newspaper when told to.

There is also reliable evidence concerning yet another poodle which can talk, this time a black female of medium size. Trudi belongs to a couple who live in Essex. So far she has learned to say clearly the words "No" and "I want one", and when her owner makes a telephone call he sometimes holds out the receiver to her; according to what Trudi hears at the other end of the line, she barks down the phone either "No" or "I want one".

The dog which holds the record in modern times for its talking prowess is a female black-and-white German sheepdog, Corinna by name. She was trained to speak by her owner, who lives in Prague, and his wife. This extraordinary dog far outshines even Don in his accomplishments. One day in 1959 Mme Tumova surprised her husband with the announcement that it seemed to her as if Corinna was trying to learn to speak, and had come out without any prompting from anyone with the word "*Mám*" (I have). After this, Gustav Tuma began to take a more lively interest in his pet. In the course of the next few months Corinna learned a second word, "*ráda*" (with pleasure), then a third, "*balona*" (ball). After half an hour's coaching every day for eleven months Corinna was able to say quite distinctly: "*Mám ráda balona!*" (= I like the ball).

Since then Corinna has developed into a speaking marvel and, what is even more surprising, has shown talents in other directions as well. Without making any mistakes she can distinguish the colours red, white, and blue. If told to fetch the blue ball, or the white or the red one, she does so without ever bringing the wrong ball back.

Besides enjoying her games with balls Corinna likes sweets. It was therefore natural as the next step to teach her the word "*bonbon*". Soon she could say "*Mám ráda bonbon!*" (I like sweets!).

Up to the summer of 1962 Corinna's vocabulary increased considerably. She was able to give correct answers to questions put to her in Czech as well as in English. If her owner told her to say hallo to a visitor she would bark out "*Dobry den!*" (Good morning); if asked her name she would give it in the Czech form, "*Corinna Tumova*"; but if, after giving her answers, she was asked what all that had to do with the visitor anyway, she would say unequivocally and forcefully "*Hovno!*" (Damn all!). If then her owner pretended to be annoyed or said in a loud voice "*Corinna, co to je?*" (Corinna, what's that?) she would again turn her head a little to one side and say quite clearly "*Hanba!*" (Disgraceful!). It is not hard to imagine the hilarity occasioned among guests by such exchanges.

Corinna has her own method of making clever use of the words "*balona*" and "*bonbon*" in other contexts. If she is hungry she goes over to her bowl and barks out the word "*bonbon*" two or three times in quick succession, so that her mistress knows that she wants something to eat. If she feels like going for a walk she fetches her lead, puts it down in front of her mistress, and says "*balona! balona!*"

For some time now Corinna has also been learning English words. She can say "How do you do" and answers questions put to her in English with the word "Well". If there is a knock at the door she calls out "Come in!" In the spring of 1962 Gustav Tuma and his wife began teaching Corinna her first words of German. She is the first animal ever to be able to express itself in three languages.

A talking poodle was the subject of a thesis for a doctorate in the first decade of the nineteenth century. The student engaged on it was considering the question how, by skilful manipulation of the larynx, deaf-mutes could be helped to produce articulated sounds. That student later became Dr Horalny, first director of the Deaf-Mutes Institution founded in Prussia. The subject of his thesis, a medium-sized black poodle, which a retired officer had taught to imitate a number of words, was once presented by its owner to the Queen of Prussia. The poodle is reported by Dr Horalny to have astonished the Queen by greeting her with the words "*Gott grüsse Euch!*" (God be with you!). Dr Horalny maintains that this poodle could bark out the alphabet from A to Z—a rather improbable feat.

Poodles have also frequently been presented in music-halls and circuses by their trainers as 'counting' dogs. From various accounts over the last two hundred years it is quite clear that poodles are far cleverer performers than any other breed of dog. In writing about 'talking' dogs it is important not to overlook those dogs which, instead of uttering more or less articulate sounds, make themselves understood by means of tapping out their answers with their paws. They are very numerous. The 'tapping' and 'counting' dogs are very quick to learn and understand, by a sensitive study of their teacher, how to interpret every gesture, even the most insignificant. But first of all we should say something about the real masters of the art of tapping and counting—horses. At the beginning of the present century they caused a sensation and gave rise to protracted discussions in the press and in exalted scientific circles as to how much reliability could be attached to their achievements in this field. Their outstanding representative was a horse named Clever Hans.

CHAPTER THREE

CLEVER HANS

THE life story of Wilhelm von Osten is that of an idealist whose hopes and dreams were thwarted by fate. The son of an aristocratic Prussian landowner, he was convinced that his horse, Clever Hans, could read and count. This black stallion created a sensation towards the end of the last century and in the first few years of this. He remained a topic of conversation right up to the beginning of the First World War.

Seeing that there were lively discussions among scientists, not only in Germany but further afield too, about the possibility of these horses having the power of independent thought, and seeing also that it is universally accepted that both von Osten and Karl Krall, who later worked with Hans and other horses, acted in good faith in their untiring efforts to teach the horses arithmetic, it would seem worth while to report on these events in some detail.

By the time the publicity for von Osten and Clever Hans had become widespread, the 'horse professor' had already developed into a crank, an old man who wore ragged clothes and a gold watch-chain with a silver horseshoe on his paunch. Not a vestige remained of the dashing young man who spent most of his time with the horses on his father's estate when, in a back-yard in Berlin, he began to teach a black stallion which he had bought in Poland to count.

Day in, day out, von Osten placed wooden boards in front of

the stallion with numbers painted in white on them. These were followed by small, thin boards each showing a letter of the alphabet. When called upon to do so, the horse pawed the ground up to five times according to which number between one and five was written on the board. Besides this, Clever Hans could be ridden without reins, and responded correctly to words of command such as "right, left, straight on, halt, slower, faster".

Von Osten, who at 28 had inherited enough money to enable him to devote himself full-time to his teaching experiments, had bought his house in Berlin as early as 1888. He purchased countless books about animals, but in addition works dealing with speech therapy for deaf-mutes and simple school textbooks. He never missed a visit to the Renz Circus or, later on, the Busch Circus. When at length he felt qualified to go ahead with his plan to teach horses he purchased a stallion descended from an Orlow trotter, with which he persevered for years without achieving any real success. When this horse, the first Clever Hans, died just before the turn of the century he bought the stallion which was later to create such a furore. He felt that this time he had struck lucky, and regarded his new acquisition as being a far more intelligent animal.

It is hardly surprising that von Osten's neighbours were eager spectators when he held his lessons in the yard behind his house. After two years' coaching he became absolutely convinced that the second Clever Hans understood a great deal of German and could also read words in German script, provided they were written with small initial letters. He believed that Hans could count up to 30 and could also understand the four main processes of arithmetic. He rated Hans's intelligence high on account also of his ability to distinguish various colours; however, as every circus trainer knows, distinguishing colours is one of the easiest tricks which horses can be taught.

Meanwhile von Osten had turned to another, highly intricate method of teaching. He set up skittles in order to explain to the horse the concept of numbers by varying the number standing. Clever Hans seemed to grasp this concept, for he counted them correctly by tapping with his right hoof. Several months later he progressed to the point of being able to specify the tens with his left hoof.

Certain now that Clever Hans had mastered not only the multiplication table up to ten but also adding and subtracting, von Osten began to teach him the alphabet according to number values. For this he made use of a phonetics board about the size of a school blackboard. It is impossible to describe here the whole complicated phonetics system used. At all events, Clever Hans proved to be very quick on the uptake, and von Osten firmly believed he had succeeded in achieving something no-one had ever done before, proving a horse capable of independent thought. Frantically he continued his tuition. The neighbours were getting annoyed with the 'mad horse-slaver' who shouted at the patient animal and even threatened it with his whip. In spite of this, the number of spectators was growing and growing, for word had got around of what was going on in von Osten's back-yard. It was well known that he never took money from spectators, besides which he had refused a number of offers to present Clever Hans at the Busch Circus and the Wintergarten in return for considerable sums of money.

The old man had other ideas: he was seeking scientific recognition. His efforts to obtain an expert opinion were at first unsuccessful; not one of his pressing invitations was answered. Downcast and disappointed, he decided to sell Clever Hans. But the advertisement which he drafted was returned to him by the newspaper on the grounds that it was a leg-pull. Finally he managed to have the following advertisement printed in a different newspaper:

> I wish to sell my 7-year-old stallion, which is as gentle as a lamb, and with which I have been conducting experiments to establish the horse's mental ability. He can distinguish ten colours, knows the four processes of arithmetic, etc.
>
> v. Osten, Berlin, Griebenowstr. 10.

As only horse-dealers and knackers applied, apart of course from a few practical jokers, von Osten made another attempt to interest men of science, and turned now to the Psychological Society, whose president, Professor Albert Moll, agreed at least to go and see Clever Hans. His opinion was a shattering blow to von Osten. He considered that the stallion provided no proof whatever of any ability to think independently nor was it in any way capable of working out any problem for itself. It had not escaped the Professor's notice that the horse reacted only to scarcely perceptible signals, just as trained circus horses do. He thanked von Osten for arranging the demonstration for his benefit, but tactfully omitted to mention that in his view the experiments were a waste of time. Von Osten inferred from Professor Moll's friendly note that he did not wish expressly to give due recognition to the successful teaching experiment, no doubt for reasons of envy, and proceeded to invite anyone who cared to come along to see the horse perform.

At this point reporters appeared on the scene to form their impressions as to whether this horse having lessons in a yard crowded with spectators really was a 'thinking' horse or whether the old man's claims about Clever Hans were the product of a deranged mind. The press opinions were by no means unanimous; three leading newspapers described the horse's performance unequivocally as deception, and one went so far as to sum up the whole affair as "humbug". On the other hand, there were quite a few popular newspapers, as well as a number of serious journals, which were of the opinion that the animal's achievements were nothing short of amazing. Well-known personalities of the day wrote dramatic reports about the 'wonder horse' which found their way into newspapers abroad. To these people Clever Hans was a rewarding topic in more senses than one, but for his owner there was nothing but expense and irritation.

Von Osten remained unshakeable in his belief in Clever Hans's capabilities, and went so far as to petition the Emperor Wilhelm II to order a scientific investigation into the horse's mental ability. The Emperor passed on the petition together with his recommendation to the Prussian Ministry of Culture, as a direct result of which two eminent scientists were sent along to von Osten's Berlin address. One of them summed up his achievements as "the most astounding thing which has ever happened in the field of animal pedagogics, an achievement which is of major significance to Science".

On August 12th, 1904, Dr Studt, the Minister of Culture, himself went to visit von Osten and Clever Hans, taking with him two other professors of science in addition to several high-ranking army officers and the governor of the province of Brandenburg. This impressive body of scientists and dignitaries caused the editors of several of the leading newspapers to despatch their top reporters to the scene of the official investigation. Perhaps, the sceptics began saying, perhaps there is something in all this after all. . . . And the rumour was circulating that His Majesty himself was desirous of seeing the horse and its owner soon, a rumour which was publicized as far afield as the U.S.A., where it was reported in the *New York Times*.

Dr Studt was soon firmly convinced that Clever Hans could not only do addition and subtraction but could also give the numerator and denominator when confronted with problems involving fractions, as is clear from a report dated the day after the visit. The amazing horse seemed to be further advanced mathematically than many twelve-year-old schoolboys, judging by the following report which appeared in the *Staatsbürger-Zeitung*:

> And so the birthdays of the Emperor and his lady wife, and of His Highness the Crown Prince, as well as the days which commemorate Sedan and Christmas, were given with a precision which caused great astonishment. Minister Studt gave expression to his unqualified admiration for the capabilities of the horse that reads and counts.

A few days after the visit of the Minister of Culture there arrived the Emperor's personal aide-de-camp, Major-General von Scholl, followed a week later by Count Moltke, to report back to the Emperor in even greater detail on this marvel among horses.

It is very probable that the Emperor would have been pleased either to summon von Osten and Clever Hans to his palace or to call on them in person, but there was now increasing clamour in the press denouncing the whole affair as "sheer folly", "a very fishy business", and "stuff and nonsense", and the Emperor therefore deemed it more prudent to await the outcome of the report then being prepared by a scientific commission.

Another supporter of Clever Hans and his owner was the African explorer C. G. Schillings, who in a series of lectures claimed for the horse "a degree of intelligence never before attained by any other horse". He persuaded the highly sceptical director of the Zoological Gardens, Dr Ludwig Heck, to visit the horse. His finding was that while he was full of praise for the horse's capacity for memorizing he did not believe that it was the result of his own thinking, but merely the successful consequence of persistent training.

Finally, on September 6th, the learned scientists paid von Osten the visit he had so long pressed for. The commission comprised Professor Nagel, head of the department of Physiology at Berlin University; Dr C. Stumpf, director of the Psychological Institute, member of the Academy of Sciences, university professor and Privy Councillor; a major-general, a veterinary surgeon, and an inspector of schools; the circus owner Paul Busch; and the Zoo director with his assistant.

After an investigation lasting several days, this august body reached the conclusion that neither the giving of signs nor tricks of the kind commonly taught such animals would account for what they had seen. In a guarded statement released on September 12th, 1904 and signed by all the members of the commission it was revealed that "certain of the undersigned have become acquainted with the method of Herr von Osten, which is essentially different from that of animal training and is modelled on primary-school instruction procedure". It ended by announcing "an earnest and thorough scientific investigation".

Were the experts going to produce proof that Clever Hans really was a 'thinking' horse? Could von Osten now proudly announce to the world that the experts had given the sought-for recognition to his successful teaching method?

The foremost authorities on animals at that time, the circus owner Paul Busch and the circus director Albert Schumann, declined to ascribe to Clever Hans any unusual capacity for thought. They were firm in their conviction that it was all connected with some effect on the horse's hearing or sight the details of which had still to be ascertained. It was known that horses had been trained along such lines and exhibited for centuries at fairs and market-places. This view was, however, rejected out of hand by the majority of the Berliners who had seen Clever Hans. The controversy raged even more fiercely. Meanwhile there appeared on the scene several 'counting' horses which were undoubtedly very intelligent, and also one dog skilled in the art, presented to the public by none other than the artist Rendich, who had written an article on Clever Hans. All these animals obeyed certain specified signs indiscernible to the onlooker. This prompted the scientific commission to test Clever Hans exhaustively and to put von Osten severely to the test too. Four times a week Professor Stumpf, Oskar Pfungst, and the recording clerk were taken by coach to von Osten's address. Simple and not-so-simple sums were given to the horse to solve by tapping with its hooves; it was given questions to answer; and it was fitted with blinkers to prevent it from reacting to any signs which might possibly be made to it while the questioner was standing to one side of it or behind it. When von Osten became offended and refused to bring the horse out for further experiments, Oskar Pfungst began a series of experiments independently.

The inevitable happened: on December 9th, 1904 the opinion of Professor Stumpf was published in the daily newspapers; the judgement on Clever Hans was pronounced, and included the following extract:

> The horse refuses when the solution to the problem set is unknown to any of those present. It can therefore not count, read, or reckon. It also refuses when it is sufficiently hampered by large blinkers from seeing the persons to whom the solution to the problem is known, especially the questioner. It therefore requires visual assistance. This assistance need not be deliberately given. Proof of this lies in the fact that the horse gave correct answers to a greater number of persons when Herr von Osten was absent. The horse must, in the course of its long period of instruction, have learned to take note, during its pawing of the ground in its responses, to an ever increasing degree of accuracy, of the small alterations in the posture of the body which the teacher unconsciously makes when reaching the conclusion of his own thought processes and which serve as signals to the horse to stop. The motivation for the channelling of its exertions and attentiveness in these directions lay in the reward which it regularly received afterwards in the form of carrots and bread.

In conclusion the report credited Herr von Osten with having acted in all good faith, and reiterated that in spite of the grave self-deception of this extraordinarily patient man, the painstaking care he had taken was not worthless to Science, since it had been demonstrated thereby that "daily perseverance over a period of four years had produced in the horse no trace of conceptual thinking".

Von Osten, who felt humiliated, gave up any further attempts to teach Clever Hans and in the summer of 1907 retired to his country estate. But he did not take Clever Hans with him; he left him in his stable in the care of a master carpenter named Piehl.

Meanwhile there had appeared a new teacher for Clever Hans, a jeweller named Karl Krall, who came from Elberfeld and had no doubts whatever about the mental potential of horses. Between 1905 and 1908 he gave Clever Hans several thousand lessons, asserting with great conviction in his book *Thinking Animals* (1912): "The horse works just as well with blinkers as without them, and in darkness just as well as in the light. The only explanation possible to justify the total achievement of the horse is that it is correct to assume the horse capable of independent mental activity."

Every few weeks Krall travelled from Elberfeld to Berlin to continue his instruction. In May 1909 he visited von Osten on his estate. Now 71, von Osten was suffering from cancer of the liver. He no longer felt kindly disposed towards Clever Hans—still unable to appreciate that it was he himself who had erred. About this Krall wrote: "His last words to me were a curse on Clever Hans, whom he blamed for the bitter misfortune of his own life. His persistent and deep hatred of his horse, which he would have liked to see run down by a heavy lorry, lasted to the end of his days."

On June 29th, 1909 the would-be discoverer of the horse capable of abstract thought died, disillusioned and alone. Clever Hans, for his part, did not suffer the fate wished upon him by his embittered owner, but was taken into the care of Krall, who sent him off for a six weeks' summer holiday on a private estate in the country, after which he was taken to Krall's large, clean stable in Elberfeld.

CHAPTER FOUR

KARL KRALL'S SECRET

In the stable at Elberfeld Clever Hans found himself in the company of two 'mentally strong' Arab stallions, Muhamed and Zarif, which Krall had bought from a stud in association with Major-General Eugen Zobel, who remained firm in his belief that horses are capable of abstract thought.

Every day since their arrival in November 1908 Krall had been teaching them to read and count, using a system similar to that of von Osten. Krall, however, believed that his methods were an improvement on those of his former colleague. Instead of tapping with their hooves directly on to the paved floor of their thatched 'classroom', they were taught to tap, at a given signal from Krall, on a counting board made of soft beechwood set at an angle. Krall had noticed in Berlin that Clever Hans, by tapping hundreds of thousands of times on the pavement of the courtyard behind von Osten's house, had developed a hoof disease—which had, however, been cured by the carpenter into whose care Clever Hans had been placed after von Osten refused to have anything further to do with the horse.

Krall formed the impression that his two Arab stallions possessed greater mental faculties than Clever Hans and that his teaching method produced palpably better results. Krall was convinced that Muhamed and Zarif "recognized numbers both when they were spoken out loud and when they were chalked up on the board".

With the precision of a schoolmaster Krall wrote letters and numbers on the board fixed to the wall. He also possessed a clock suitable for teaching the time, about the size of clocks found on the walls in banks, with hands which could be moved about easily. Other visual teaching aids were a wall calendar with slots into which numbers could be fitted and a simplified alphabetical table by means of which the horses were meant to learn to relate the numbers which they heard with the corresponding numbers which were pointed out to them visually. On signals given by Krall, Muhamed and Zarif tapped out answers to questions put to them which to all appearances were phonetically correct. It was already being said that horses were capable of being taught to perform certain tricks from a limited range of words of command which had, however, to be reinforced by gestures. But Krall firmly believed that his horses recognized people from photographs and were able to identify them by name or by some description of them. Thus he claimed to have taught them to react to a photograph of the Kaiser (Emperor) Wilhelm II by tapping out 'keiser', a phonetically correct version. Muhamed and Zarif are also said to have recognized portraits of the poet Schiller and the philosopher Schopenhauer and to have spelt out their names with hoof-taps. In the case of Schopenhauer the result was rather less successful because, as Krall explained, the horses could do no more than suggest the phonetic sequence of the name with the corresponding number of hoof-taps.

The experiments went even further. Krall plagued his patient pupils with lessons in which he taught them French words, and in his explanations of the supposedly multilingual learning capacity of his horses went into long-winded comparisons with the teaching of languages to children. He deluded himself into believing that the horses could tap out '*dö*' (for *deux*) just as well as '*zwei*' (= two) and that they had mastered arithmetical problems in the French language involving numbers up to 100. Krall even maintained that his pupils had grasped the fact that in French, unlike German, the tens precede the units—thus '*vingt-quatre*' (twenty-four) but '*vierundzwanzig*' (four-and-twenty).

Everything so far described was, however, mere child's-play

compared with the 'independent thought' utterances of the horses and their regular conversations with human beings. Krall interpreted the hoof-taps by reference to the special alphabet on the wall and so transformed the sequence of letters into words. The horses also addressed him by name. At least, they tapped out '*kral*' or '*krl*'. Requests for food they spelt out with the words '*Brot*' or '*Brt*' (bread), '*Mören*' or '*Müren*' (carrots), '*Zukr*' or '*Zuker*' (sugar). When they were in a bad humour they simply refused to answer or to give any 'independent utterances'.

Now, well over half a century later, one can only admire the patience of the jeweller and his friends who participated in the education of these horses. He believed in all seriousness that the expert opinion of the animal psychologist Oskar Pfungst in 1904 was wrong and that Clever Hans was in fact quite capable of reading and calculating. Pfungst had, however, proceeded very conscientiously and had carried out protracted experiments with twenty-five horses "to ascertain whether the creatures really understand human speech or whether they merely react to the sound and intonation of voices and likewise to the expressive movements and involuntary gestures which inevitably accompany speech". In his book *Herr von Osten's Horse* (1907) Pfungst came to the conclusion that horses definitely do not understand human speech as such, but do react to the expressive movements, however slight, of humans. Horses are in this respect like dogs, he claimed, excellent 'muscle readers'.

Krall surprised everybody shortly afterwards by announcing that he was not confining his instruction in reading and arithmetic to Muhamed and Zarif, but was purchasing other horses in order to train a whole class of 'thinking' horses. In addition he bought a pony, which he named Hänschen "in honour of Clever Hans", and, with a view to invalidating Pfungst's assertion that horses reacted only to signs, he bought a blind cart-horse from Pomerania, named Berto.

Krall and his friends were so certain that they had opened up entirely new paths which would lead to a closer understanding between men and animals that in 1913, with Professor H. E. Ziegler, they founded a 'Society for Animal Psychology', which issued its own news-sheet. This survived until 1934—several years after the death of Krall, which occurred in 1929. A small publisher in Stuttgart who brought out this little-sought-after news-sheet was ruined, having also financed various books on 'reading' and 'calculating' dogs, which sold hardly any copies.

Serious scientists dissociated themselves from the very outset from this pseudo-scientific 'Society for Animal Psychology'. Even so, a number of eminent men of science who had achieved distinction in their various fields of research, including Professor Claparède, did contribute articles, though they soon discovered that they were treading on dangerous ground. Time has shown that their articles contributed nothing worth while to the cause of serious research into the field of animal behaviour.

Krall continued his instruction in reading and calculating with horses after the outbreak of the First World War and until the middle of the following year, when through pressure from high government authority, he was obliged to call a halt. The Commission responsible for supplying horses to the Cavalry refused exemption even to 'thinking' horses, and apart from the pony Hänschen and the blind cart-horse Berto all the horses (Krall owned ten 'pupils') were drafted into military service. They never returned to Elberfeld.

Suffering from a heart complaint, Krall moved after the war to Munich. In 1927 he gave his circle of friends further food for thought by declaring himself to be a believer in thought transference between men and animals.

In his book *Thinking Animals*, which appeared in 1912, Krall included a coded entry concerning his work:

> Inferences to be Drawn from My Work
>
> For the protection of my rights I submit in the following form a number of results which form the basis of my further experiments:
> aaaaaaaaaaaaaaaa bbbbbb cccc ddddddddddddddddd
> eeeeeeeeeeeeeeeeeeeeeeeeeeeeeeeeeeeeeeeeeeeeeeeeee ffff
> ggggggggggggggg hhhhhhh iiiiiiiiiiiiiiiii kkkkk lll
> mmm nnnnnnnnnnnnnnnnnnnnnnnnnnnnnnnnnnnnn o
> pp rrrrrrrrrrrrr sssssssssssss ttttttttttttt
> uuuuuuuuuuuuu vv ww zz

For fifteen years he kept to himself the secret of this entry. Whether it was known to Professors Claparède, Kraemer and Ziegler would seem in retrospect doubtful. It was not until 1927, when Krall published, in the form of a supplement to an issue of the news-sheet for the 'Society for Animal Psychology', a booklet entitled *Thought Transference between Men and Animals*, that his supporters and friends became acquainted with his 'confession of faith'. "The intimate understanding horses have among themselves", he declared, "is based upon extra-sensory thought transference. Thought transference can also exist between men and animals and has nothing to do with the independent capacity for thought of the recipient."

Karl Krall died on January 12th, 1929, in the firm conviction that he had had genuine conversations with horses and had taught them to solve square roots and cube roots.

CHAPTER FIVE

THEY 'TALK' AND 'COUNT' WITH THEIR PAWS

THE number of 'tapping' and 'counting' dogs which have been able to tune in to the wave-length of their human trainers with excellent results is very great. They all react to the slightest signs which, however, remain undetected by even the sharpest eyes in an audience. The story of these dogs could fill a whole book.

There were, for example, the 'reading' and 'counting' poodles exhibited by a certain Andreas Pschorrer in the seventeenth century. "Look at this gentleman closely, Hasso, and tell me how old you would judge him to be," he would say, whereupon the poodle would pick out from a number of boards each displaying a different number the one which it 'thought' to be the right one. Another poodle would invariably retrieve objects which its master had distributed among members of the audience who then concealed them—a trick still performed by 'thinking' dogs. Pschorrer later presented his dogs to Louis XIV at Lyons, and afterwards to William III, King of England, at Amsterdam; by this time his 'reading' poodles had learned a number of new tricks, and he had added to his repertoire 'counting' terriers and dancing pomeranians. Several years later, shortly after the Turks had been forced to retreat from the gates of Vienna, Pschorrer's 'thinking' dogs were presented at the court of the German Emperor Leopold I, who demanded to have the training methods explained to him in detail and had the cleverest of the poodles repeat their best tricks, expressing afterwards his admiration for their feats of memory.

At the time of Frederick the Great there were a number of dogs well known for their counting skill. A Flemish dog-trainer who travelled Europe under the name Aigle de Cir was much applauded for his troop of 'counting' dogs, the stars of which were Berry, a Scottish sheepdog, and Loney, an English terrier.

From this period also there are reports about a dog which was being exhibited in 1754 in the city of Danzig. This animal, when asked the colour of a dress, would select the correct board from a number each painted with a different colour, or when asked the time would pick the right card from among twelve numbered 1–12. He would also answer certain questions, with a 'yes' or 'no' by tapping with his right forepaw, and solve simple arithmetical problems, apparently without assistance, again by tapping with his paw.

From about this time any number of 'wonder dogs' have appeared on the scene, answering questions by tapping, and solving problems involving long division, multiplication, and subtraction—seemingly only after careful thought—for audiences at music-halls and circuses.

A big sensation was caused in New York and other American cities during the first half of the nineteenth century by a 'thinking' dog called Munito. He played cards at table with his master, turning the cards over with his paw. Besides this he could select from a variety of wooden boards, with letters of the alphabet painted on them, ones which joined up to form words. Also, by tapping with his paws, Munito could give the correct answers to arithmetical problems, and even 'answered' questions put to him by the audience.

In the present century exceptional interest has been aroused by certain 'thinking' dogs which, in the opinion of their owners, were so clever that it was possible to converse with them just as if they were human beings. These 'wonder dogs' caused nearly as great a stir as the 'thinking' horses. Few of them have become as famous as the airedale Rolf and a bitch descended from him, the paw-tapping Lola. Rolf, the subject of much controversy in 1913, belonged to a certain Paula Moeckel from Mannheim.

Apparently he was not only able, with the help of a simple tapping alphabet, to answer questions but could also follow conversations. It was therefore possible to 'converse' with Rolf.

Frau Moeckel claimed that she noticed quite by chance that Rolf could understand what people were saying. One day she had been helping her daughter Frieda with a fairly simple addition she was having difficulty with. She noticed that Rolf was behaving in a strange manner, demanding attention by placing his paw several times on her knee. Jokingly she asked Rolf whether he knew the answer to the problem. To test him she asked, "What's two plus two?", whereupon Rolf unhesitatingly tapped her arm lightly four times with his paw. Flabbergasted, she then asked him, "And what's five plus five?", at which Rolf tapped ten times. A tall story perhaps, but people believed it.

Frau Moeckel then began to teach Rolf in the following way: she 'settled' with him the number of taps to be used to represent the letters 'a', 'e', 'i', 'o', 'u', etc.; two taps signified the letter 'o', three 'a', ten 'e', thirteen 'i', eighteen 'u', while eight taps represented 'r' and six 'w'. Rolf appears to have expressed his words phonetically, tapping out at times a 'k' for a 'g' or an 'f' for a 'v', according to how these letters are pronounced in German.

For a time the newspapers and magazines were full of articles on Rolf. Frau Moeckel wrote a whole book on him, with a complete record of the 'utterances' made by Rolf independently and without any assistance, and revealed that he spelt out his own name as Lol. According to her, Rolf seemed to prefer expressing himself in the Mannheim dialect—hardly surprising, considering that he heard it every day wherever he went.

Several scholars, including Professor Ziegler, were frequent callers at Frau Moeckel's house and got to know Rolf well. Like Professors Gruber and Kraemer and Dr Sarasin, he came to the conclusion that there was no form of deception involved and furthermore that Frau Moeckel was quite right in all the assertions she had made.

Other well-known people, such as Dr Wilhelm Neumann, the author of works on animal psychology, reached entirely opposite conclusions. In his book *Man and Beast* Dr Neumann reported a number of details from his own 'conversations' with the 'thinking' dog Rolf which do not in any way appear to justify the claim that the dog was capable of independent thought. He even mentioned that for an airedale—a particularly intelligent breed—Rolf ought rather to qualify as mentally retarded, besides which he had some nasty habits: "As a result of all the titbits brought to him by the numerous visitors he was grossly overweight. I personally observed that, though four years old, he was still not house-trained."

Eight times Dr Neumann had 'conferences', each lasting for several hours, with Rolf, at which he made in all twenty-three experiments in an attempt to achieve successful results with at least one of the much-vaunted 'conversations'. As Frau Moeckel had died a short while previously, some of these sessions were attended by her three daughters, while at others there were no eyewitnesses.

The results were surprising. Rolf, it was revealed, only gave the right answers when a member of the family was present. If, apart from this, he was asked questions to which none of the members of the family present knew the right answer Rolf too was unable to answer or reacted by giving utterly confused tapping signals. Moreover, if Rolf was shown objects without their being seen by a member of the family and he was told to let them know in a neighbouring room what object was being held up to him, he merely tapped out one of the following 'dilemma phrases': "*Gehn lassen!*" (Leave me alone), "*Buckel steigen!*" (= Go and jump in the lake), "*Mag nit*" (Don't want to), "*Lol krabbelt nit auf Neumann seinen Leim!*" (Lol (=Rolf) is not going to fall into Neumann's trap).

Dr Neumann comments plainly and sarcastically: "The 'dilemma phrases' originated in the dilemma in which the Moeckel daughters found themselves because they did not know which answer to make the dog tap out." In short, Dr Neumann came to much the same conclusion as Dr Pfungst had ten years earlier with Clever Hans: "The dog tapped out purely mechanically, without having any inner participation in the matter, what the member of the family wanted it to."

However, various dogs which Professor Ziegler knew, as well as their owners (whom, one feels, the professor knew rather better), appreciated him as an expert, and some of the dogs even 'dictated' letters to him. It makes one shake one's head to learn that his own dog Awa was supposed to have dictated a letter to the jeweller Krall informing him that Professor Claparède from Geneva had visited his friend Professor Ziegler. In those days such assertions from a scholar carried a good deal of weight, but time has effaced the errors of men like Claparède, Krall, and Ziegler. Dr Neumann comments:

> Anyone reading Ziegler's treatises on 'paw-tapping' animals will notice that time and again, with typical emotionalism, he makes the same unsubstantiated propositions and illogical inferences which give a highly uncharacteristic and regrettable bias to the later years of this zoologist who in other respects is so deserving of praise.

The years which followed came up with hordes of 'counting' and 'talking' airedales, fox-terriers, pinschers, poodles, sheep-dogs, wire-haired terriers, spaniels, pomeranians, and dachshunds. The majority of these had lady owners who were firmly convinced that their darling pets could think for themselves because they tapped with their paws or barked. One striking aspect of all this is that almost invariably some authority could be found to provide an 'expert' opinion, yet the real authorities on such matters—the professional trainers, who publicly presented dogs which they had taught to perform in just such a way—were never approached.

The owners of such 'wonder dogs' derived a great deal of enjoyment from the astonishment of guests whom they invited to watch their pets 'thinking'.

It was by no means unknown for 'reading' and 'counting' dogs to lose the thread of their public performances and create incidents, some of them hilarious. On one occasion a dog which was supposed to be playing the piano allowed itself to be lured away by a sausage thrown on to the stage by one of the spectators. To the intense amusement of the audience the piano went on playing just the same. When the dog, oblivious of the fact that it had unmasked its trainer, resumed its place at the piano the audience applauded and laughed and whistled to such a pitch that the terrified creature sought refuge with its master, while the piano continued to play the same tune.

Another dog, this one a 'counting' dog, ruined a circus performance all because a lady in the audience had brought along with her a small female puppy in her bag. Completely overcome by its tender feelings for the sweet little puppy, the four-legged virtuoso forgot its act.

At a circus in 1950 an amusing incident occurred during a performance at which I was present. Karl Faszini had just begun his dog revue act when a cat came rushing into the tent hotly pursued by Roland, the circus owner's bulldog. The terrified cat leapt into the ring, whereupon the poodles, terriers, and pomeranians jumped down from their stools yapping furiously and chased it right across the ring and under the spectators' seats, where the bulldog got wedged and had to be pulled out backwards. The revue was considerably delayed before the performing dogs could be put back on their stools.

These examples clearly show how little even well-trained dogs 'think' when their concentration is interrupted by some sudden, startling incident.

CHAPTER SIX

DOG DETECTIVES

It is truly amazing what dogs are capable of learning. For intelligence and quickness in learning, scientific experiments and animal trainers alike place them second only to monkeys. One undoubtedly significant factor is that, after thousands of years of close contact, dogs have become thoroughly used to their human masters and are therefore able to guess instinctively what is required of them, even before any words of command are uttered. Just as nowadays they fetch letters and newspapers and slippers to their master, so in ancient Greece and Rome they fetched papyri and writing tablets and soft leather house shoes.

Dogs are quick to learn to differentiate orders. Fetching and carrying belong just as much to their natural gifts as jumping over obstacles. Yet there are still quite a number of canine talents which remain unexplained. Homer speaks of their good powers of memory, and Aristotle too, in his work *Parva Naturalia*, attempts to explain the memory of dogs. Even in those days it was well known that it was the emergence of memory images which caused the vivid dreams, sometimes accompanied by vocal utterances, which dogs have.

Hans Thum, the first president of the International Poodle Breeders' Society and inventor of the caracal cut for poodles, states in his book *My Friend the Poodle* (1961) that in their contact with humans dogs have extended their innate use of smell and hearing by exploiting another sense, that of sight. In addition to the success already achieved with dogs in 'reading' and 'counting' he mentions the possibility of thought transference and commands given telepathically, but rather inclines to the view that in such matters success is due to signals, given consciously or unconsciously. Thum refers to the fact that experienced hunters regard telepathic transmission of commands as being quite feasible. The fact that dogs hurry to the door at the approach of their master, even if he is still a long way off and they can neither see, hear, nor scent him, would seem to indicate a psychical contact between dog and master, "a proof of its extremely refined sensory perception, which far exceeds human receptivity". Thum adds that the harmony which exists between man and dog further increases the psychic contact—a fact which will be confirmed by every dog owner who knows how to treat his dog properly. Examples are given of better-than-average performances by intelligent dogs, which could be explained as manifestations of their loyalty, allied to their natural gift of sagacity.

In the fourteenth century a greyhound which avenged its master's death became so famous that a statue was erected in its honour. On a mantelpiece in an old castle near Fontainebleau dating back to the time of the Valois King Charles V stands a statue depicting a fight between a powerfully built greyhound and a man. It commemorates the dog belonging to an aristocrat, Aubry de Montdidier, which helped to clear up the mystery of its master's murder. According to the story, Aubry de Montdidier, an officer and confidant of the King, had sent his dog Hercules to his bride-to-be to announce his impending arrival. The clever dog, who knew every inch of the way, duly arrived—but his master did not.

Sent off on the road home, Hercules found his murdered master in the forest of Bondy—now a north-east suburb of Paris—buried under some leaves. For three days he kept guard over the body, then ran off in search of the young sieur de Narsac, Aubry's closest friend. Seeing Hercules looking forlorn and half-starved, de Narsac fed him. Hercules then made it clear, by tugging at de Narsac's doublet, that he should accompany him, which de Narsac did, accompanied by two armed companions. They followed Hercules across Paris in a north-easterly direction till they reached the forest of Bondy, where they found Aubry. From that moment on Hercules refused to leave his master's friend.

A few days later de Narsac, accompanied by Hercules, happened to meet the knight Macaire, one of the crossbow archers of the King's bodyguard. Hercules immediately attacked the knight, and tried to seize him by the throat. This caused Macaire to be strongly suspected of the young Count Aubry's murder. Until that time Hercules had been extremely friendly towards all soldiers.

Naturally word got round about the dog's strange behaviour towards Macaire. King Charles heard about it. As he had thought very highly of Aubry, he took it upon himself to investigate the case. He paraded all twenty of his noble crossbowmen in front of the Saint-Pol palace, his Paris residence, and Hercules was then brought on to the parade ground. Again he attacked Macaire. When the King questioned him Macaire denied having killed Aubry, even though evidence of a quarrel between them had meanwhile come to light. The King decided to submit Macaire to 'ordeal of battle'—an ancient practice whereby the suspected person was obliged to fight anyone who charged him with guilt, in this case the greyhound Hercules. The place chosen for the battle was the Ile de la Cité—which at that time was still uninhabited—in the middle of the river Seine. Macaire, who took up his position at one end of an arena specially marked out for the contest, with Hercules at the other, was allowed to use only a cudgel and a shield against his opponent. When the signal was given for battle to commence, Hercules rushed at Macaire, seized him by the throat and knocked him to the ground. Mortally afraid, Macaire screamed: "Yes, I did it—just take the dog away and I'll tell you everything!"

After Macaire had confessed to the murder of Aubry through jealousy of the King's favour towards him, Charles had him beheaded. As for Hercules, a memorial stone recounting his deeds was erected to him in the forest of Bondy. Strolling-players and minstrels sang the dog's praises, and in 1816 the actor Karsten wrote a melodrama entitled *Aubry's Dog or The Forest of Bondy*.

For the part of Hercules he trained a royal poodle, and travelled with his troupe and its star member from one theatre to another. It was through this poodle that the poet Goethe ceased to be director of the Weimar Court Theatre, for he refused to allow a dog to perform on his stage, whereas the Grand Duke of Weimer, his patron, was very keen to see the dog act its part in the play. The performance took place, and Goethe left for Jena.

About 1780 a poodle named Mignon, belonging to Monsieur Carpillon, the bell-ringer of Notre-Dame, brought to light a terrible crime. This servant of the church, who was held in the highest regard by everyone, suddenly disappeared without trace. The police established that his poodle, which had been handed over to the man's next of kin, always became very restless when taken past the shop of a barber named Galipaud on the corner of the rue Marmouset and the rue des deux Ermites. Monsieur Carpillon had been a regular customer of Galipaud and had met many of his acquaintances there. It was then discovered that a number of other customers of Galipaud had also disappeared mysteriously, and the police had been unable to trace any of them.

Two police officers were detailed to accompany Mignon to the barber's shop. There, when they let him off the lead, he whined and scratched at a door which led the police into a room at the back of the courtyard, where they discovered a number of human bones in a tub, together with an axe and some knives of the type used to slaughter animals.

As soon as the barber was fetched from the shop by the police the poodle attacked him furiously. Galipaud was accused of being a multiple murderer and, in the face of the incontrovertible evidence, he confessed that he had killed a number of people, dissected their bodies, and extracted certain organs which had been passed on to the owner of a small restaurant nearby, an Italian pastry-cook named Grimandi. The bell-ringer Carpillon was his last victim. After a sensational trial Galipaud and Grimandi, whose meat pasties had been eaten by hundreds of his unsuspecting customers, were sentenced to be beheaded.

These extraordinary events from the fourteenth and eighteenth centuries could be matched by many others. Long before dogs were officially brought into police service these four-legged amateur detectives were used in solving crimes.

One astounding and often puzzling feature of dogs, cats, and horses is their uncanny sense of direction, which enables them to find their way back home from almost anywhere. They can return from distances of hundreds of miles even if they have not made the outward journey on foot but by car or rail. There are any number of stories to illustrate this remarkable faculty, which have never been satisfactorily explained.

An almost unbelievable, but well-authenticated, story concerns a poodle named Moffin which accompanied his owner, a Milanese corporal in Napoleon's army, everywhere on his campaigns, including the Moscow campaign. In the chaos caused by the retreat from Moscow, the corporal lost Moffin while crossing the Beresina in November 1812. Almost a year later there appeared at the barracks in Milan a shaggy, unkempt dog which ran straight past the sentries on duty, just as though it knew its way round the place. Hurrying on through the corridors, Moffin found his old master, whom he greeted by jumping up at him and barking. At first the corporal failed to recognize the dog as his, but when he did there was a very touching reunion. This devoted poodle had found his way back from beyond the Alps over a distance of more than 1,250 miles. At first it was thought that Moffin had made friends with another soldier and so was brought back, but after extensive enquiries this was shown not to be the case.

In 1872, according to an equally well-authenticated source, a dog found its way back home over a distance of more than 900 miles. In Menton, on the French Riviera, a man named Milandri, the proprietor of the Hôtel Victoria, owned a silky pomeranian which he sold one day to the Archduchess Marie Rainer, who happened to be staying at the hotel. On her return to Vienna she took the dog with her, but before long it had run away. Months later it turned up in an unkempt and emaciated condition on the Riviera to rejoin its master. The small dog had taken the arduous route across the Alps with unerring precision, as though it possessed an inbuilt compass.

In his *History of the Dog* the French author Blaze writes

about the devotion of dogs, justly known as man's best friends, and their love of home.

> The greatest happiness of a dog consists in remaining in the proximity of its master. Should the latter be obliged to beg his bread, it will not merely assist him in this important undertaking, but will not desert him, even if a king himself were prepared to accept it into his royal residence. Friends will desert a man in his misfortune, but his dog will remain faithful to him for ever.

Every dog can be taught to behave in a certain way and to perform tricks. These do not necessarly have to be connected with reading or counting. Dogs can be trained, for instance, not to jump up at their master to show their joy; they can be taught not to bark, or to obey the commands 'Sit!' or 'Lie down!' indoors just as quickly as they learn the meaning of 'Heel!' when out walking. The motorist can accustom his dog to occupy a certain place in the car, usually in the back. The cyclist can train his dog to keep to the near side when running alongside, to prevent danger from other road traffic.

Almost all house dogs can distinguish between commands like 'Fetch the slippers' and 'Bring the newspaper' and can learn to obey such commands without error. And they are quick to learn to jump over a stick or other obstacles placed in their path. Nearly every intelligent dog needs hardly any training to sit up and beg, provided it afterwards receives the reward which it feels it has earned. It is comparatively easy to teach it to progress from sitting up and begging to walking a few steps in an upright position. Another, but more difficult, trick is to turn its body in a 'pirouette' while walking upright, and many house dogs never accomplish this—which is not surprising, as such a trick really belongs to the more sophisticated repertoire of circus-trained dogs. For such tricks the advice of a professional trainer should be sought.

Specialized training is needed for guide dogs for the blind and for police dogs. Dogs selected for these purposes often show themselves to be exceptionally intelligent. While a blind person makes use of his dog's seeing eyes, thus shielding himself from the dangers of traffic and collision with pedestrians and obstacles, the police make use primarily of its sense of smell, in order to follow scents.

The guide dog has to know all the obstacles which could be a danger to its master, such as level-crossing gates, pillar-boxes, and so on. It is therefore trained in a mobile harness. Lengthy training is needed for it to be able to cope successfully with such procedures as stopping at the kerb-side before crossing the road and negotiating flights of steps. As most dogs used by the police are concerned with bringing criminals to book, only strongly built dogs are eligible, belonging to such breeds as airedales, boxers, dobermanns, rottweilers, and alsatians, besides of course powerful mongrels. They are trained in special police-dog schools.

The importance of the alsatian in police work is discussed by Dr Ursula Siers-Roth in a book entitled *Das Constanze-Hundebuch*:

> In 1949 the plain-clothes police force [in Germany] investigating burglaries and thefts by breaking and entering made 1,088 arrests with the help of alsatians. During the same period police dogs tracked down 80 murderers (in England and Wales 119 and in the Benelux countries as many as 181).

Over 95% of the dogs used on the Continent by frontier police are alsatians, according to the same writer. Though airedales and rottweilers have been more extensively used since then for such work, there is still a preponderance of alsatians.

In the history of the plain-clothes police forces of Germany, England, and France the greatest successes achieved so far have been due to the efforts of alsations. The main reason for this is of course that plain-clothes police, railway police, and frontier police prefer using alsatians because they show themselves so adaptable during training. But, over the past fifty years, the rottweiler has proved to be a close second in every European country.

Ulrich Klever, in *Knaurs Hundebuch*, writes:

> Two great police dogs were a rottweiler from Kiel named Rex and a tracker dog belonging to the London police named Ben. Ben tracked down over a hundred criminals while on active service, while Rex earned for himself fame and publicity for the part he played in the arrest of fourteen brawling sailors in Kiel in 1912.

Dr Siers-Roth reports further on the Kiel incident that Prince Henry of Prussia, brother of the Emperor Wilhelm II and very popular in his role as 'Navy-Prince', had difficulty congratulating Rex's handler personally. The handler had first to obtain the Prince's permission to remove his helmet, "as this was the signal to the dog that it must behave civilly and that there was no occasion for arresting His Highness . . . !"

PART FIVE
THE FRIENDLY DOLPHINS

CHAPTER ONE

THE RIDE TO SCHOOL

EVEN in ancient times, observers were puzzled by the playful and friendly ways of dolphins. Pliny and Plutarch wrote detailed accounts of these sea creatures sacred to Poseidon. What Pliny, in particular, reported on their activities along the French Mediterranean coast, and in the Bay of Puteoli, where they are said to have been especially docile, was long regarded as imaginative fiction.

In recent decades, however, there has been a change of attitude, resulting from the publication of observations made by American research workers on dolphins in Florida and on one of the islands in the Gulf of Mexico and of others made on tame dolphins in biological marine stations in Japan. Besides these, authentic stories about these affectionate and intelligent creatures have been received from coastal areas of New Zealand.

As a result, the accusations levelled against Pliny have been withdrawn, and the modern reader is astonished at what Plutarch wrote, about A.D. 100, on the relationship between these smallest of all cetaceans and man: "The dolphin is the only creature which attaches itself to man out of pure affection. Dogs and horses are docile mainly because they receive their food from man. Only the dolphin has been granted the divine gift of unselfish friendship, dreamed of by the noblest thinkers."

Many fishermen and shore-dwellers of antiquity prized the

friendship of these elegant sea-mammals. In the Middle Ages too, sailors of the Mediterranean and the Atlantic coastal waters of France were close observers of dolphins. Their descriptions of them found their way into the courts of princes and from 1340 the kings of France accorded the title "Dauphin", the French form of the word 'dolphin', to their heirs apparent.

Many surviving coins from ancient Greek and Roman cities show a boy riding on a dolphin. For centuries there was a relief depicting a boy on a dolphin in Poseidon's temple at Taenarum, on the southern tip of the Peleponnese. This sculpture, carved by order of Periander, ruler of Corinth, is said to commemorate the miraculous experience of his friend the celebrated poet and minstrel Arion of Lesbos. As Periander ruled from 627 till about 587 B.C., we know roughly the decade in which this strange event took place.

With the full approval of the Prince of Corinth, Arion had travelled to Italy and Sicily to sing to his own harp accompaniment. He had been showered with gifts and was now returning on a Corinthian vessel. On the way he was robbed by the avaricious captain and his crew and had to jump overboard to avoid being murdered. The story is graphically recounted by Herodotus:

> Arion dressed himself in the full costume of his calling, took his harp, and standing on the quarter-deck, chanted a lively air pitched in a high key. His strain ended, he flung himself, fully attired as he was, headlong into the sea. The Corinthains then sailed on to Corinth. As for Arion, a dolphin, they say, took him upon his back and carried him to Taenarum.

The rescued minstrel arrived in Corinth before the boat and went immediately to tell the prince what had happened. When the sailors went ashore they were arrested and convicted of robbery and attempted murder.

The significant words in the account of Herodotus are "pitched in a high key". Through modern research we know nowadays that dolphins taught to mimic human words do in fact reproduce them in a high-pitched register since the dolphins' own vocalizations are falsetto, so we can say that Arion's ride on a dolphin is entirely credible. It would seem that the dolphin, attracted by the sound of the harp, was already tracking the boat when the minstrel leapt from it and, looking on him as a playfellow, carried him safely on its back to the shore.

The Romans occupying the province of Gallia Narbonensis, in southern Gaul, were well acquainted with the methods used by the fishermen along the coast to catch large quantities of fish with the help of dolphins. These methods must have been studied at first hand by Pliny, as he was Roman tax-collector in Gallia Narbonensis for a number of years about A.D. 70. In Book IX of his *Natural History* (Heinemann: Loeb Classical Library) he writes:

> At a regular season a countless shoal of mullet rushes out of the narrow mouth of the marsh into the sea, after watching for the turn of the tide, which makes is impossible for nets to be spread across the channel—indeed the nets would be equally incapable of standing the mass of the weight even if the craft of the fish did not watch for the opportunity. For a similar reason they make straight out into the deep water produced by the neighbouring eddies, and hasten to escape from the only place suitable for setting nets. When this is observed by the fishermen . . . and when the entire population from the shore shouts as loud as it can, calling for 'Snub-nose' for the dénouement of the show, the dolphins quickly hear their wishes if a northerly breeze carries the shout out to sea, though if the wind is in the south, against the sound, it carries it more slowly; but then too they suddenly hasten to the spot, in order to give their aid. Their line of battle comes into view, and at once deploys in the place where they are to join battle; they bar the passage on the side of the sea and drive the scared mullet into the shallows. Then the fishermen put their nets round them and lift them out of the water with forks. None the less the pace of some mullets leaps over the obstacles; but these are caught by the dolphins, which are satisfied for the time being with merely having killed them, postponing a meal till victory is won. . . . When the catch has been completed they tear in pieces the fish that they have killed.

A little further on, Pliny reports that the dolphins, "aware that they have had too strenuous a task for only a single day's pay, wait till the following day, and are given a feed of bread mash dipped in wine, in addition to the fish."

Pliny describes these events not only to provide evidence of the dolphin's intelligence but also to explain how mullet, regarded in Rome as a great delicacy, was caught. A basket (50 Kg.) of these red fish, which measure about a foot in length, and weigh up to a kilogram (just over 2 lb.), would be bought by Roman epicures for 5000–8000 sesterces—a handsome sum, considering that a slave, as Martial tells us, cost 1300 sesterces.

A similar procedure practised by dolphins in the Gulf of Mexico to catch flying-fish is described by the nineteenth-century American zoologist Audubon. There the "gauchos of the sea", as he calls them, form a team which drives the flying-fish and grey mullet on until they die from exhaustion. Fishermen in the Gulf of Mexico noticed that the concerted efforts of the dolphins often resulted in large shoals of fish being systematically driven into lagoons where the largest and tastiest could be selected at leisure.

The following story of the boy who rode to school on a dolphin in the Bay of Naples did most to give Pliny the undeserved reputation of being a teller of tall stories.

Every day a fisherman's son had a four-mile walk from Baiae to Puteoli to get to school. Today children get there by rail or bus, but in those days they had to follow the path skirting the Lucrine Lake, unless they were taken by boat.

All the fishermen in the bay were on good terms with the dolphins there. Far from hunting them, they fed them. So it is not surprising that the boy of whom Pliny writes had become friendly with them too. In fact, he had made a special friend of one particular dolphin, with which he swam and played. When the dolphin was not to be seen from the shore the boy would call it in a high-pitched voice: "*Simo, Simo*" (Snub-nose), whereupon, "although it was concealed in hiding it used to fly to him out of the depth". The dolphin would often let him sit on its back and then take him for a ride round the bay.

Making the most of the opportunity, the boy steered his willing companion across the bay to where his school was. The dolphin quickly accustomed itself to the route, and also to returning at the right time to take the boy back home, waiting just off shore till the boy's call was heard.

Apart from this story and the one told by Herodotus about Arion, there are a number of others which describe tame dolphins that have come down to us from writers of antiquity. Pliny the Younger has written at great length about the dolphin of Hippo. Hippo Regius, an old Carthaginian settlement, used to stand on what is now the East Algerian town of Bone, and there too a dolphin became friendly with a fisherman's son.

Furthermore, a hundred years later Oppian, in a poem describing how fish were caught off the island of Lesbos, tells of the friendship between a dolphin and a fisherman's son who used his friend to take him for rides out to sea. This same poet also mentions the way dolphins and fishermen worked together in the bay between Euboea and Attica. Similarly in the Bay of Iasion, on the Mediterranean coast of what is now Turkey, two tame dolphins are said to have given rides to young boys.

In ancient times dolphins were not only represented on coins and statues. Because of their speed they formed part of the motif on the columns along the track used by the athletes in the Circus Maximus, in Rome.

However well the sailors of every period of history may have known dolphins, the scholars of medieval times and even recent times have been very ill informed. The great naturalist Buffon wrote: "It is maintained that porpoises, sea-calves and dolphins approach such vessels as make resonant music on board in still weather; but this fact, about which I have doubts, is not related by any writer of note."

Well over a century ago Professor P. Scheitlin noted that dolphins have excellent hearing and are capable of producing sounds: "They call to one another and engage after a fashion in human speech. They are said to have a remarkable attachment to humans. It has always been asserted that they like music and greatly enjoy singing and the sound of the harp, thereby being attracted to ships."

CHAPTER TWO

PELORUS JACK AND OPO

IN the nineteenth century a number of stories, confirmed by numerous eyewitnesses, were being told about the clever ways of dolphins. There was, for instance, the story of the dolphin off the New Zealand coast which acted as pilot to ships negotiating the hazards of Cook Strait, which separates North Island from South Island and is treacherous in places because of submerged rocks. Between 1888 and 1912, whenever a steamer approached Pelorus Sound a dolphin would appear ahead of it and accompany it through the channel. No matter whether the steamer was sailing from east to west or west to east, the "swimming pilot" would always be there, and after circling the vessel would move into position ahead and adjust itself to its speed. All the sailors and passengers who used this route were delighted to see their "pilot", whom they called Pelorus Jack. Sailing ships were of no interest to him, and so it may be assumed that it was the noise of the engines that attracted him to steamers.

In 1904 the Governor of New Zealand passed a protective order, under which anyone found hunting dolphins was liable to a fine of £100. This order was renewed several times, but in spite of it Pelorus Jack disappeared without trace just about the time of a voyage through the Strait by four Norwegian whalers. It was never established whether he had been caught by the Norwegians or whether he had fallen victim to a killer whale, which are known to include dolphins among the sea creatures they kill for food.

Round the coasts of New Zealand fishermen's children used to make friends with the affectionate, friendly dolphins. In the shallow bays they would splash about with them, knowing that while these creatures were there they were safe from sharks—for dolphins will not tolerate sharks anywhere near them. If a shark comes too close (unless it is one of the very large species, which in any case do not venture inshore) they form groups and drive it off.

Between January and March 1955 (which is summer-time in New Zealand), the Auckland, Christchurch, and Wellington newspapers were full of stories of the games children were playing with a dolphin in Hokianga Harbour.

Reporters for the *New Zealand Herald* and the *Evening Post* described the relationship of mutual trust which had developed between Jill Baker, a thirteen-year-old New Zealander at Opononi, and a tame dolphin. They were quick to point out the similarities between what was happening locally and Pliny's account of the fisherman's son in the Bay of Puteoli. Jill and her friends not only played water polo with their pet, but also went for rides on his back. Jill, a good swimmer, rode out to sea on him, and was always brought safely back to shore.

When Opo, as the dolphin came to be called, began putting in regular appearances the north-east corner of North Island became a busy excursion centre for the inhabitants of Auckland, the largest town in New Zealand, situated to the south-east, and Wellington, the capital, at the southern tip of the island. Every week-end during the twelve months of Opo's daily appearances in Hokianga Harbour there were traffic jams stretching all the way along the coast road. Many thousands of New Zealanders, besides visitors from Australia, wanted to see this amazing spectacle for themselves. They were eyewitnesses to one of the most important events in the history of wild-life study, but more than this they regarded Opo's appearance in the midst of crowds of holiday-makers enjoying a bathe as a sign of providence. In their enthusiasm some of them tried to stroke Opo, who enjoyed the mass adulation and willingly submitted to the treatment handed out to him.

A protective law was passed for Opo too. On the very day it was passed, however, Opo suffered a tragic fate: he became wedged between protruding ridges of rock and perished when the tide went out. And so this lovable creature passed from the scene.

Japanese newspapers, reporting these events at the time, supplemented them with authentic stories from round the coasts of Japan. In various fishing villages on the main island of Hondo and on Kyushu, to the south, children had made friends with dolphins. The newspapers noted too that in several of the oceanaria, built on the pattern of the research station in Florida, even more astonishing facts were coming to light. They also reported an event which had taken place during the Second World War and was no doubt amusing to the Japanese, though not to those directly involved. Some American airmen, shot down by the Japs, had taken to their rubber dinghies. The Japs, out of range of their guns, were able to watch the airmen's desperate efforts to ward off a 'helpful' dolphin which wanted to push the boat precisely in the direction of the Jap-occupied island! The dolphin was not to know, of course, how unwelcome its efforts to help the Americans were. It must have regarded their behaviour as rank ingratitude—for even the friendliest of dolphins can turn bitter when all they receive for their pains is to be struck with paddles. . . .

Both on the Californian coast and off Florida, as well as along the coastline of Hokkaido, dolphins have been observed irresistibly pushing back to shore bathers who have swum too far out to sea. The explanation for this behaviour may be traced in the characteristic peculiarity some dolphins have of playfully pushing with their heads large pieces of driftwood and buoys that have come adrift from their moorings. Sometimes these creatures have actually been credited with saving the lives of exhausted swimmers. It would, however, be going too far to suggest that such rescue acts are performed deliberately in order to save the lives of those in peril.

CHAPTER THREE

PROGRESS IN RESEARCH

In the Marineland oceanarium on the Florida peninsula, efforts have been made, with considerable success, over the past few decades to tame dolphins and teach them to perform tricks.

This oceanarium, situated in St Augustine, near Daytona Beach on the east coast, directly facing the Atlantic Ocean, was established in 1938 as a research and observation centre. A second "dolphin studio" was later set up on the Californian Pacific coast.

Soon after the Second World War a programme was drawn up to train dolphins in the oceanarium on the Florida coast. It was decided that the best prospects for achieving this object lay in the appointment of someone with wide experience of sea-lion training, and accordingly the German animal trainer Adolf Frohn was given the job.

He carried out the first experiments in the large rectangular tank, some 330 ft. long, but soon discovered that he could establish quicker and more lasting contact with the dolphins by using the smaller, circular tank, with a circumference of about 250 ft. After a short while Frohn discovered that one of the dolphins, called Flippy, was especially amenable to training. He followed the methods of animal trainers who work in a ring with horses or camels, and at his suggestion a "dolphin ring" was built—a tank 26 ft. in diameter and just over 6 ft. deep. Here the training was carried out till Frohn could at last transfer his promising pupil into the large circular tank. From then on Flippy's tricks were the delight of many thousands of spectators daily. Later other dolphins trained by Frohn were also put on public view.

When Frohn, an excellent swimmer and diver, held out a tyre under water Flippy would swim through it. At a given signal Flippy would raise himself out of the water and catch hold of a rope in his mouth so that a bell rang, and he could also make an old motor-horn sound with a rubber ball. He understood a number of words in English, and at the word of command would leap high into the air and glide through a rubber tyre suspended about 4½ ft. above the water or through a paper-covered wooden frame.

Flippy's most remarkable feat consisted in his being harnessed to a flat-bottomed boat containing a man and pulling it across the tank, which was separated from the open sea by walls.

Considering that there is a constant supply of well over ten thousand fish of all kinds both in the rectangular and circular tanks, the tame dolphins are never short of food. Yet in spite of this they always respond to their keeper's bell signal which tells them that an extra-special morsel awaits them.

For several years dolphins have been studied in the Communication Research Institute on the island of St Thomas, which lies east of Puerto Rico and forms part of the American Virgin Islands. Here, since 1959, the neuro-physiologist, linguist and marine biologist John C. Lilly has been carrying out a series of sensational experiments on the level of intelligence and powers of vocalization of dolphins. The findings are truly remarkable.

By making use of hydrophones (highly sensitive underwater microphones) and tape-recordings, and transmitting words and phrases into the water by means of underwater loud-speakers, Dr Lilly has been able to establish that some of his dolphins mimicked them with phonetic accuracy in a high-pitched register, rather as budgerigars do. He went so far as to claim in certain learned scientific journals that with his most intelligent pupils he could conduct real 'conversations'. He had begun his experiments in the oceanarium in Florida, confining himself there to the exchange of whistling signals, with results comparable to those which can be obtained from various species of birds, such as mynahs and shama thrushes; even jackdaws, sometimes after just a few hours of listening, can repeat whistling sounds in the same pitch and with the same modulations as those they hear, and because of their powers of memory they can be taught to imitate whole melodies.

Dr Lilly, together with other marine biologists and with the dolphin keepers, went on to make a startling observation. One of the dolphins, when given a whistling signal, replied in a pitch which rose higher and higher, until finally its whistles were no longer audible. Again a parallel may be drawn with birds: only part of the song of the black-headed mannikin is audible, the rest of it becoming ultrasonic. Man can only hear sounds below a frequency of 18 kilocycles per second. As people grow older the auditory threshold recedes. Fifty-year-olds are only aware of frequencies ranging from 10 to 12 kilocycles per second, and the range of sixty-year-olds scarcely exceeds 8 kilocycles per second.

However, whereas a bird could not remotely be expected to abandon the ultrasonic range into which its song passes so as to make it audible to the human ear by using lower frequencies, the dolphins, realizing that they were no longer being understood, chose lower frequencies of their own accord.

This discovery seems so incredible that it still requires careful verification, presupposing as it does extensive thinking power on the part of dolphins. But a number of famous zoologists are in full agreement that dolphins do in fact possess an unusually highly developed faculty of thought, for not only is the weight of their brain considerable, but furthermore the number of convolutions in the cerebral cortex is high.

Comparisons, though of course they cannot represent a yardstick of intelligence, may be drawn from the following synopsis:

Just as among birds which imitate human sounds there are good and less gifted pupils, so too among dolphins there are variations in the levels of attainment. From the results of experimental work already carried out Dr Lilly knew that dolphins,

	Body weight in grammes	*Brain matter in grammes*
Child, aged 6	15,000	1000
Man, fully grown	80,000	1400–1450
Chimpanzee	80,000	350– 450
Gorilla	200,000	450– 600
Sheepdog	30,000	200– 250
Budgerigar	35	1– 2
Indian Grackle	80–100	2– 2·5
Elephant, fully grown	4,000,000	2000–3000
Dolphin	150,000–180,000	1500–1700

like other cetaceans as well as numerous species of fish, not only hear with their ears but possess a sensitive receiving apparatus for sound-waves over the whole surface of their skin. The sensitivity of their skin enables them to locate the echo of their own sonar transmissions in the ultrasonic range, which vary between 80 and 200 kilocycles per second (as is the case with certain species of bats). By means of this echo all cetaceans can determine whether there are shoals of fish or merely rocks ahead. Dr Lilly also knew from previous experiments with dolphins that they utilize the upper surface of their bodies for feeling things and enjoy contact with their human handlers, swimming beside them and rubbing against them like a dog that wants to be scratched.

With this knowledge, supplemented by his own observations arising from the acoustic exchanges with dolphins in Marineland, Dr Lilly proceeded to carry out similar experiments at the Communication Research Institute in Charlotte Amalie, the only town on the island of St Thomas, situated close to Nazareth Bay. There he found that two of the dolphins were quick to learn words spoken by him or his wife and to repeat them clearly. Even swear-words which he let slip were repeated by the dolphins, just as mynah birds and budgerigars acquire them. They imitated laughter even better than the highly gifted African grey parrot Jako could.

Dr Lilly was able to train his dolphins so well that they learned to fetch floating balls and rings at a given signal and to bounce a ball up out of the water and balance it, as sea-lions do at the circus.

The basis of training dolphins rests of course on their playful disposition and lack of hostility, which have been common knowledge for thousands of years. But successful training requires the use of stored data—*i.e.*, feats of memory—on their part. What though of the dolphin's gift of vocalization? At first one is tempted to think purely in terms of the instinct of mimicry. But Dr Lilly believes in a much more extensive intelligence so far as dolphins are concerned—to the point, in fact, of it being possible to conduct real conversations with them. With this in view, further vocalization experiments are being carried out by him with his cleverest dolphins.

It remains to be seen whether the efforts of this marine biologist, who not only speaks with dolphins but also claims that guinea-pigs imitate the pitch of human voices, will be successful.

It is quite certain that one day other exceptionally amiable dolphins like the one from the Lucrine Lake or like Opo from New Zealand will make their presence known to us. And other highly gifted dogs, like the sheepdog Corinna, will inspire the admiration of their owners and many besides for their powers of speech. Tame lions and tigers will fill their human audiences with amazement.

The behaviour of monkeys has proved in the twentieth century that they derive great pleasure from 'aping' and have a high learning potential. So many sensational results have been achieved with the study of animal behaviour in our time that further surprising revelations are bound to come.

INDEX

Index

Index